Walid Kaakeh

Agricultural Project Management and Feasibility Study

Walid Kaakeh

Agricultural Project Management and Feasibility Study

Comprehensive Guide to Planning, Evaluating, and Implementing Successful Agricultural Ventures

Noor Publishing

Cover image: www.ingimage.com

Publisher:
Noor Publishing
is a trademark of
Dodo Books Indian Ocean Ltd. and OmniScriptum S.R.L publishing group

120 High Road, East Finchley, London, N2 9ED, United Kingdom
Str. Armeneasca 28/1, office 1, Chisinau MD-2012, Republic of Moldova, Europe
Managing Directors: Ieva Konstantinova, Victoria Ursu
info@omniscriptum.com

Printed at: see last page
ISBN: 978-620-8-87056-0

Agricultural Project Management and Feasibility Study

Comprehensive Guide to Planning, Evaluating, and Implementing Successful Agricultural Ventures

By

Prof. Dr. Walid Kaakeh

International Consultant

First Edition - 2025

بسم الله الرحمن الرحيم

IN THE NAME OF ALLAH

THE MOST BLESSED, THE MOST MERCIFUL

Dedication

To my mother and father,
Aisha and Abdul Ghani Kaakeh
May God bless them and make them dwell in
His spacious gardens

To my wife Fatima Khatib

To my beloved children,
Tasneem, Yasmine, Yamen, and Enas

I dedicate this book

Walid Kaakeh

Table of Contents

Introduction

Agriculture is the backbone of many economies around the world, especially in developing countries where the sector provides livelihoods for millions of people. As the global population continues to rise, the demand for food, fiber, and other agricultural products is set to increase significantly. However, the challenges facing the agricultural sector are complex and multifaceted. These include climate change, resource depletion, market volatility, and the need for sustainable practices. To navigate these challenges and ensure the success of agricultural ventures, proper project management is essential. This is where Agricultural Project Management becomes crucial.

Agricultural project management is a multidisciplinary approach that involves planning, implementing, monitoring, and evaluating agricultural projects to ensure they meet specific objectives within time, cost, and quality constraints. Whether it's a small-scale farming initiative, a large agribusiness venture, or a government-supported agricultural development program, effective project management is essential for optimizing resources, minimizing risks, and achieving desired outcomes. The success of agricultural projects often depends on a blend of factors such as technology, market trends, financial planning, risk assessment, and environmental sustainability.

A central element of agricultural project management is the feasibility study. A feasibility study is an essential process that evaluates the viability of a project before it is undertaken. It involves analyzing market conditions, technical requirements, financial considerations, and social and environmental impacts to determine whether the project is worth pursuing. Feasibility studies are particularly vital in agriculture, where conditions can be unpredictable, and investments are often substantial. The feasibility study helps identify risks early, assess financial viability, and ensure the project aligns with long-term sustainability goals.

This book, *"Agricultural Project Management and Feasibility Study,"* serves as a comprehensive guide to understanding the key principles and best practices in managing agricultural projects, from inception to completion. It also delves deeply into the significance of conducting detailed feasibility studies to ensure the success of agricultural ventures. The book provides a robust framework for managing agricultural projects across diverse contexts, from smallholder farming initiatives to large-scale agribusiness ventures.

In Part 1 (Chapters 1-4), we explore the Introduction to Agricultural Project Management. This section provides a solid foundation for understanding agricultural projects, including their types, importance, and the various stakeholders involved. We discuss the different stages of the project life cycle in agriculture, highlighting the distinct management approaches required for small-scale versus large-scale projects. An in-depth understanding of the project life cycle enables project managers to anticipate challenges, plan effectively, and ensure smoother execution.

Part 2 (Chapters 5-8) focuses on Feasibility Study for Agricultural Projects. Here, we break down the critical components of a good feasibility study. The chapter examines the four core areas of feasibility: market feasibility, technical feasibility, financial feasibility, and social and environmental feasibility. For each of these aspects, we discuss the tools, techniques, and methodologies that can be applied to assess the feasibility of an agricultural project. The importance of conducting a detailed and well-rounded feasibility study cannot be overstated, as it provides a solid foundation for decision-making, resource allocation, and risk management.

In Part 3 (Chapters 9-11), we turn to Planning and Implementation of Agricultural Projects. This section focuses on how to develop an effective project plan, design a feasible project structure, and manage resources efficiently. It delves into key topics such as setting clear project objectives, developing a work breakdown structure (WBS), and conducting a risk assessment. The planning phase sets the stage for the successful implementation of the project. Effective management of resources—such as land, labor, and inputs—is essential for the timely delivery of agricultural projects. We also explore monitoring and evaluation techniques, discussing how to track progress, assess performance, and adapt plans as needed.

Part 4 (Chapters 12-14) presents Case Studies and Best Practices in agricultural project management. Real-world case studies provide valuable insights into the successes and challenges faced by agricultural projects around the world. We examine examples of smallholder farmer cooperatives, commercial agribusiness ventures, and sustainable farming practices, highlighting lessons learned and identifying common pitfalls to avoid. This section serves as a source of inspiration for those involved in agricultural projects, showing how careful planning, innovative strategies, and effective management can lead to success.

Also in Part 4, I reflect on the future of agricultural project management. The agricultural landscape is undergoing rapid transformation, driven by new technologies, climate change, and evolving market dynamics. I discuss emerging trends such as digital agriculture, precision farming, and climate-smart agriculture, emphasizing the need for agricultural project managers to stay ahead of these trends. The future of agriculture requires a focus on sustainability, inclusivity, and the application of cutting-edge technologies to enhance productivity and resilience.

This book is not just a resource for project managers but also for anyone involved in the agricultural sector—whether farmers, investors, policy-makers, or academics. It equips readers with the knowledge and tools necessary to understand and navigate the complexities of agricultural project management. The practical advice, real-world examples, and in-depth discussions will empower project managers to build better, more sustainable agricultural projects that contribute to food security, economic development, and environmental sustainability.

In conclusion, agricultural project management is a critical skill in an increasingly complex and interdependent world. This book provides the essential knowledge, tools, and frameworks for anyone aiming to lead agricultural projects successfully. By integrating project management principles with detailed feasibility studies, we can ensure that agricultural ventures are not only profitable but also environmentally sustainable and socially responsible.

Author
Prof. Dr. Walid Kaakeh
2025

Part I: Foundations of Agricultural Project Management

Focus: Concepts, principles, and planning frameworks.

Chapter 1: Introduction to Agricultural Project Management

Benefits:

Students and researchers gain foundational knowledge about the scope and types of agricultural projects, helping them choose a specific area (e.g., crop, livestock, agribusiness) for study or research. It also contextualizes agriculture's role in socioeconomic development, aligning research with global development goals.

1.1 Introduction

Agriculture has always played a pivotal role in human civilization, providing the food, fiber, and raw materials necessary for survival and development. With the global population rising and climate change threatening food security, agriculture has evolved into a strategic sector requiring professional planning and execution. In this context, agricultural projects have emerged as key instruments for modernizing agriculture, improving livelihoods, and ensuring sustainable development. This chapter introduces the fundamental concepts of agricultural projects, outlines their types and importance, identifies key stakeholders, and discusses the differences between small-scale and large-scale projects.

1.2 Definition of Agricultural Projects

An **agricultural project** is a time-bound, goal-oriented undertaking that involves the allocation of resources (such as land, labor, capital, and technology) to produce agricultural goods or services with a defined purpose. Projects may range from crop cultivation and livestock farming to agro-processing and rural infrastructure development.

Definition:

An agricultural project is a structured set of activities designed to achieve specific agricultural objectives within a specified timeframe and budget, often aimed at increasing productivity, income, sustainability, or social impact.

Agricultural projects differ from ongoing agricultural operations in that they are **temporary, planned, and goal-specific**, often requiring feasibility studies, budgeting, and detailed implementation plans.

1.3 Importance of Agricultural Projects

Agricultural projects contribute significantly to:

- Food Security: Projects help improve food production and distribution.
- Income Generation: They create employment opportunities and improve livelihoods.
- Rural Development: Infrastructure and capacity-building components promote rural progress.
- Innovation and Technology Transfer: Projects often introduce modern techniques, equipment, and practices.
- Environmental Protection: Projects can focus on sustainability and conservation efforts.

Well-managed agricultural projects play a strategic role in transforming subsistence agriculture into commercial, profitable ventures while enhancing resilience against climate and market risks.

1.4 Types of Agricultural Projects

Agricultural projects vary widely in purpose, scale, and implementation approach. They can be broadly categorized based on their core objectives—such as increasing production, improving sustainability, enhancing infrastructure, or building human capacity. Understanding the types of agricultural projects is essential for effective planning, implementation, and assessment.

1.4.1 Crop Production Projects

These projects are centered on cultivating crops for food, feed, fiber, and fuel. They form the foundation of agricultural activity and are often targeted at improving productivity, diversifying crops, or adopting climate-smart farming practices.

- **Examples:** Cereal farming (e.g., rice, maize, wheat), vegetable gardening (e.g., tomatoes, leafy greens), horticulture (e.g., fruit orchards), organic farming, greenhouse cultivation.
- **Key Objectives:**
    - Enhancing crop yields through improved seeds, fertilizers, and farming techniques.
    - Expanding cultivated land or optimizing land use with crop rotation and intercropping.
    - Adopting environmentally sustainable practices to conserve soil and water.
    - Reducing reliance on chemical inputs through organic or integrated pest management methods.

1.4.2 Livestock and Poultry Projects

These projects are designed to promote animal-based agricultural production. They address various components of animal husbandry, including breeding, feeding, housing, healthcare, and marketing.

- **Examples:** Dairy farms, beef cattle rearing, poultry hatcheries and broilers, goat and sheep farming, piggery units, aquaculture.
- **Key Objectives:**
    - Improving livestock productivity and reproductive performance.
    - Enhancing feed efficiency through balanced nutrition and fodder management.

- Controlling diseases through vaccination, veterinary care, and biosecurity measures.
- Developing infrastructure such as sheds, cold storage, and processing units for animal products.

1.4.3 Agri-Business and Value Addition Projects

These projects go beyond primary production to focus on processing, branding, packaging, and marketing agricultural goods. They bridge the gap between farm and market, creating economic opportunities and reducing post-harvest losses.

- **Examples:** Agro-processing (e.g., tomato paste, dairy products), packaging units, cold storage chains, agro-based startups, agritourism ventures, farm-to-fork initiatives.
- **Key Objectives:**
    - Increasing farmer income by transforming raw products into market-ready goods.
    - Creating rural employment through small and medium enterprises (SMEs).
    - Encouraging innovation in food technology and supply chain management.
    - Enhancing traceability, food safety, and consumer confidence.

1.4.4 Agroforestry and Conservation Projects

These projects integrate trees and shrubs into agricultural landscapes, promoting ecological balance and long-term sustainability. They are especially relevant in regions facing land degradation, deforestation, or climate change challenges.

- **Examples:** Tree-planting campaigns, alley cropping, shelterbelts, windbreaks, watershed management, soil and water conservation programs.

- **Key Objectives:**
  - Enhancing biodiversity and creating wildlife habitats within farming systems.
  - Improving soil fertility through nitrogen-fixing trees and organic matter.
  - Reducing erosion and runoff, especially on sloped or degraded lands.
  - Strengthening resilience to climate variability by creating microclimates and carbon sinks.

1.4.5 Irrigation and Infrastructure Projects

Infrastructure is a backbone of successful agriculture. Projects under this category aim to develop essential facilities that support agricultural operations, reduce losses, and ensure timely access to inputs and markets.

- **Examples:** Canal systems, drip irrigation, boreholes, farm ponds, warehouses, silos, rural roads, market sheds.
- **Key Objectives:**
  - Ensuring reliable water supply for irrigation, especially in arid and semi-arid areas.
  - Reducing dependency on rain-fed agriculture through modern irrigation systems.
  - Minimizing post-harvest losses via cold storage and improved transportation.
  - Enhancing rural connectivity and market access to boost profitability.

1.4.6 Research, Training, and Capacity-Building Projects

Research, training, and capacity-building projects are fundamental to the long-term transformation and sustainability of agriculture. Unlike production- or infrastructure-focused initiatives, these projects center around people and knowledge systems, aiming to empower individuals, institutions, and communities with the skills, tools, and innovations needed to thrive in an evolving agricultural landscape.

In a sector increasingly shaped by climate variability, technological change, globalization, and policy shifts, building human capacity is not just a complementary strategy—it is a cornerstone of successful agricultural development.

Purpose and Importance

These projects serve multiple strategic functions:

- Bridging the knowledge gap between research institutions and farmers.
- Facilitating the dissemination and adoption of new technologies, such as improved seed varieties, precision farming tools, or organic pest management techniques.
- Empowering rural communities with education, skills training, and leadership development.
- Strengthening institutions and governance systems that oversee agricultural innovation and rural extension.

In many developing regions, low productivity and resource inefficiency are closely linked to the lack of access to timely, accurate, and locally relevant information. These projects fill that void, offering the support mechanisms needed to drive inclusive growth and sustainable practices.

Examples of Projects

1. **Agricultural Extension Services**

   Government or NGO-run programs that provide technical advice, problem-solving support, and updated farming knowledge to smallholder farmers via field agents.

2. **Farmer Field Schools (FFS)**

Group-based experiential learning where farmers learn by doing—observing, experimenting, and evaluating new methods in real field settings.

3. **Climate-Smart Agriculture Training**
   Programs focused on helping farmers adapt to climate change through knowledge of drought-resistant crops, agroecology, soil conservation, water harvesting, and crop diversification.
4. **Digital Advisory Platforms**
   Mobile apps and SMS services offering weather forecasts, market prices, planting tips, pest alerts, and real-time agronomic advice, often in local languages.
5. **On-Farm Research Trials**
   Collaborative experiments between researchers and farmers to test innovations like seed varieties or fertilization methods under local conditions.
6. **Youth and Women Empowerment in Agribusiness**
   Incubator programs, training camps, and entrepreneurship support that enable young people and women to build viable agribusinesses or take leadership roles in value chains.
7. **Institutional Strengthening Programs**
   Projects aimed at building the capacity of universities, agricultural colleges, extension departments, and cooperatives through curriculum development, staff training, and policy support.

Key Objectives

1. Strengthening Skills of Stakeholders

These projects enhance the knowledge, technical skills, and decision-making abilities of:

- Farmers, through participatory learning, demonstrations, and peer exchanges.
- Extension agents, by equipping them with up-to-date training tools and diagnostic skills.
- Agripreneurs and rural youth, via mentorship, market literacy, and business incubation.

2. Promoting Adoption of Improved Technologies

They serve as conduits for:

- Disseminating research breakthroughs (e.g., pest-resistant crops, drip irrigation).
- Demonstrating the benefits of sustainable farming techniques (e.g., intercropping, integrated pest management).
- Reducing the innovation lag between research stations and farm fields.

3. Building Institutional Capacity

Such projects often involve:

- Upgrading research infrastructure (labs, trial plots, ICT platforms).
- Developing policy frameworks for innovation systems and knowledge sharing.
- Improving governance and accountability in agricultural organizations.

4. Enhancing Resilience and Adaptive Capacity

Through education and outreach, they aim to:

- Increase awareness of climate change risks and appropriate responses.
- Improve financial literacy, including access to credit, insurance, and farm budgeting.
- Teach risk management strategies, such as crop diversification, conservation practices, and market linkages.

Cross-Cutting Themes

- **Gender Equity**: Projects often include a focus on reaching and empowering women farmers, who are disproportionately affected by limited access to training and resources.
- **Youth Engagement**: By integrating ICT tools, agri-tech, and innovation hubs, many programs aim to make agriculture attractive and profitable for the next generation.
- **Sustainability**: These initiatives reinforce the importance of long-term environmental stewardship, social inclusivity, and economic viability.

Illustrative Case

In Kenya, the "*Digital Green*" initiative collaborates with local extension systems to deliver video-based agricultural training to rural farmers. This has increased technology adoption rates by over 30%, reduced training costs, and reached tens of thousands of smallholders—demonstrating the power of combining digital tools with grassroots capacity-building.

Conclusion

Each type of agricultural project plays a unique role in the broader goal of achieving food security, economic growth, and environmental sustainability. While some projects are production-oriented, others focus on enabling infrastructure, market integration, or human capacity. Understanding these categories allows stakeholders to design more effective, tailored interventions that address specific agricultural challenges and opportunities. By integrating technical innovation with community participation and environmental stewardship, agricultural projects can drive inclusive and transformative change across rural and urban landscapes alike.

Table Summary of Agricultural Project Types

A clean, professional table summarizing the categories, focus areas, and sample projects.

Project Type	Primary Focus	Examples
Crop Production Projects	Increasing yields, optimizing land use	Cereal farming, horticulture, organic farming
Livestock & Poultry Projects	Animal care, breeding, disease control	Dairy farms, goat farming, poultry hatchery
Agri-Business & Value Addition	Processing, marketing, agritourism	Food packaging, cold storage, farm branding
Agroforestry & Conservation	Sustainability, erosion control, biodiversity	Tree planting, shelterbelts, watershed mgmt
Irrigation & Infrastructure	Water supply, storage, connectivity	Drip irrigation, silos, rural roads
Research & Capacity Building	Training, technology transfer, farmer empowerment	Extension services, field schools, ICT tools

1.5 Key Stakeholders in Agricultural Projects

Successful agricultural projects involve a wide range of stakeholders:

Stakeholder	Role
Farmers and Producer Groups	Primary beneficiaries and implementers
Government and Policy Makers	Provide regulations, subsidies, and support
NGOs and Development Agencies	Support planning, funding, and monitoring
Investors and Financial Institutions	Provide capital and risk-sharing instruments
Extension Officers and Technicians	Offer technical training and advice
Agro-Input Suppliers	Supply seeds, fertilizers, tools, and machinery
Markets and Retailers	Provide the channels for selling the produce

Each stakeholder plays a crucial role in the design, implementation, monitoring, and sustainability of the project.

1.6 Agricultural Project Life Cycle

Agricultural projects, like all development or business initiatives, go through a systematic series of phases known as the project life cycle. This structure ensures that the project is conceived, planned, executed, and evaluated in a logical and efficient manner. However, agricultural projects have unique characteristics—such as seasonality, biological variability, and environmental dependencies—that require this life cycle to be adapted to the agricultural context. Understanding each phase is crucial for students, practitioners, and policymakers involved in agricultural development.

1.6.1 Initiation Phase

The initiation phase marks the beginning of the agricultural project. It involves:

- **Identifying Needs and Opportunities**: This step entails recognizing the problem that needs to be addressed—such as low yields, market failures, or food insecurity—or spotting a viable opportunity, such as expanding organic farming or introducing a new value chain.
- **Understanding Constraints**: Agricultural projects are influenced by environmental, socio-economic, institutional, and policy factors. Identifying these constraints early (e.g., land tenure issues, climate risks, input shortages) helps avoid major bottlenecks later.
- **Preliminary Assessments**: These may include basic technical, financial, and social assessments to gauge viability.
- **Stakeholder Consultations**: Farmers, local communities, government agencies, NGOs, and other stakeholders are consulted to ensure inclusivity, community ownership, and relevance.

Example: Before starting a dairy development project, local assessments may reveal insufficient veterinary services and poor road infrastructure—insights critical for shaping the project.

1.6.2 Planning Phase

This phase converts ideas and assessments into a **practical roadmap**. It is the most critical stage for long-term success. Key components include:

- **Detailed Feasibility Studies**: These involve technical feasibility (e.g., soil suitability for crops), financial and economic analysis (cost-benefit ratios, IRR, NPV), social and environmental impact assessments.
- **Objective Setting**: Clear, measurable objectives are defined (e.g., increase maize yields by 20% in three years).
- **Timeline Development**: This includes seasonal considerations, such as planting and harvest cycles.
- **Resource Planning**: Budgeting, financing options, manpower needs, land, equipment, and input requirements are detailed.
- **Risk Management Plans**: Includes identifying weather risks, pest outbreaks, or market fluctuations and planning mitigation measures.

Importance: Effective planning sets the foundation for achieving desired outcomes while minimizing risks and resource wastage.

1.6.3 Execution Phase

This is the **implementation stage** where planned activities are put into action. It involves:

- **Procurement of Inputs**: Seeds, fertilizers, equipment, and other agricultural inputs are sourced and distributed.
- **Hiring and Capacity Building**: Staff and field personnel are recruited and trained. Farmer training, demonstration plots, and extension services are implemented.

- **Field-Level Implementation**: Land preparation, sowing, irrigation, livestock care, pest control, harvesting, and other field activities are carried out as per the work plan.
- **Coordination and Supervision**: Timely coordination among stakeholders ensures alignment and smooth operations.

Example: In a rice intensification project, this phase would involve training farmers on new planting methods, distributing improved seed varieties, and coordinating irrigation schedules.

1.6.4 Monitoring and Control Phase

No agricultural project can succeed without regular performance checks. This phase focuses on **tracking progress, ensuring quality, and making real-time adjustments.**

- **Performance Monitoring**: Defined indicators (e.g., plant survival rate, input usage efficiency, market prices) are used to monitor progress.
- **Data Collection and Analysis**: Field officers and monitoring teams gather data using surveys, reports, satellite images, or mobile apps.
- **Corrective Actions**: If actual performance deviates from plans, adaptive measures are taken. For example, pest outbreaks may prompt emergency pesticide applications or a change in crop protection strategy.
- **Financial Monitoring**: Ensures expenditures are within budget and funds are properly utilized.

Value: This phase reduces waste, improves efficiency, and builds accountability across all project components.

1.6.5 Closure and Evaluation Phase

Once all field activities are complete, the project enters its final phase. Closure and evaluation include:

- **Final Reporting**: Comprehensive documentation of activities, expenditures, and outcomes. This includes production reports, financial statements, and lessons learned.
- **Impact Evaluation**: Measures the effectiveness of the project against its original goals. Evaluations may be internal or done by third parties and can include qualitative and quantitative assessments.
- **Stakeholder Debriefing**: Sharing outcomes with beneficiaries, policymakers, and funding agencies helps ensure transparency and learning.
- **Sustainability Planning**: Identifying ways to maintain project benefits post-funding, such as community ownership, integration with local governance structures, or private sector partnerships.

Example: A vegetable farming initiative may be evaluated based on yield improvements, farmer incomes, women's participation, and reduction in food imports.

1.7 Differences Between Small-Scale and Large-Scale Agricultural Projects

Criteria	Small-Scale Projects	Large-Scale Projects
Scope	Community or individual level	Regional or national level
Investment	Low to moderate	High capital investment
Technology	Traditional or semi-modern	Mechanized, high-tech
Management	Informal, farmer-led	Formal, often professionally managed
Risk	Lower exposure, but less resilience	Higher risk, but diversified

Understanding these differences is critical when designing interventions and feasibility studies, as project size significantly affects risk profile, sustainability, and outcomes.

1.8 Summary

This chapter has laid the foundation for understanding agricultural projects by defining their nature, scope, and importance. We explored the various types of projects, the roles of different stakeholders, the life cycle of projects, and the critical distinctions between small- and large-scale operations. These concepts are essential for the effective planning and execution of agricultural projects and will be further elaborated in the following chapters, especially in the context of feasibility analysis and implementation.

Chapter 2: Key Stakeholders in Agricultural Projects

Benefit:

Understanding stakeholder roles and power dynamics allows students to analyze real-world case studies more critically and design stakeholder-inclusive projects. Researchers can also frame stakeholder analysis models or conduct participatory research.

2.1 Introduction

Stakeholders are individuals, groups, or institutions that are affected by or can affect the outcome of a project. In agricultural projects—where outcomes often intersect with social, economic, environmental, and political dimensions—stakeholder engagement becomes critical to achieving sustainable results. This chapter explores the identification, roles, and influence of stakeholders in agricultural project planning and execution, while also highlighting best practices for stakeholder analysis and involvement.

2.2 Defining Stakeholders in Agricultural Projects

Stakeholders are defined as any individuals, organizations, or entities that have an interest in or are impacted by the outcomes of a project.

Stakeholders can be direct (actively involved in the project) or indirect (affected by the project outcomes but not involved in implementation). Identifying all relevant stakeholders early in the project life cycle helps in managing expectations, allocating roles effectively, and reducing potential conflicts.

2.3 Categories of Stakeholders

Agricultural project stakeholders can be broadly grouped as follows:

2.3.1 Primary Stakeholders

These are the main beneficiaries or implementers of the project.

- **Farmers and Farmer Groups**: Individuals or cooperatives involved in production. They are the most crucial stakeholders since their adoption of practices or technologies determines success.
- **Local Communities**: The broader population in the project area that may benefit from infrastructure, training, or improved access to food.

2.3.2 Secondary Stakeholders

These are supporting actors who influence or enable the project.

- **Government Agencies**: Provide legal frameworks, subsidies, and policy support.
- **Extension Services**: Deliver technical advice and capacity-building.
- **Non-Governmental Organizations (NGOs)**: Offer field-level support, mobilization, training, and even funding.
- **Private Sector Partners**: Agro-input suppliers, buyers, traders, and processors involved in supply chains.
- **Financial Institutions**: Banks, microfinance institutions, and cooperatives that provide credit and financial products to support project activities.

2.3.3 Tertiary Stakeholders

These stakeholders **indirectly influence** the environment in which the project operates.

- **Researchers and Academics**: Provide innovations, pilot projects, and baseline data.
- **Media and Civil Society**: Help disseminate information and raise awareness.
- **International Donors and Development Agencies**: Fund projects and influence strategy through their priorities and conditionalities.

2.4 Stakeholder Roles and Responsibilities

Stakeholder	Roles and Contributions
Farmers & Cooperatives	Participate in implementation; provide local knowledge; test and adopt innovations
Government Ministries (e.g., Agriculture, Environment)	Set policy, ensure compliance, provide subsidies or technical support
NGOs	Facilitate community entry, offer training, and build capacity
Agro-input Suppliers	Provide seeds, fertilizers, equipment, and advisory services
Buyers & Processors	Create market access; provide price signals and quality standards
Financial Institutions	Offer loans, credit facilities, insurance, and investment support
Researchers	Evaluate technologies; recommend best practices; monitor impact
Media	Inform stakeholders; enhance project transparency; influence public opinion

2.5 Stakeholder Influence and Interest Matrix

One useful tool for stakeholder analysis is the Influence-Interest Matrix, which helps project planners understand the power and interest of different actors.

Influence	High Interest	Low Interest
High Influence	Key players – engage closely (e.g., farmers, funders)	Keep satisfied (e.g., regulators, policymakers)
Low Influence	Keep informed (e.g., local NGOs, community groups)	Monitor (e.g., media, distant consumers)

Tip: Prioritize engaging stakeholders with high interest and high influence early in the project.

2.6 Stakeholder Engagement Strategies

Stakeholder engagement is a critical success factor in agricultural project management. It goes beyond simple communication or awareness campaigns—effective engagement is a dynamic, ongoing, and iterative process that spans the entire project cycle, from identification to closure.

Engaging stakeholders ensures that diverse perspectives are considered, promotes ownership, enhances social equity, and reduces resistance. Here are key strategies to ensure meaningful and productive engagement:

1. Mapping and Prioritizing Stakeholders

Before engaging stakeholders, it's vital to identify, categorize, and prioritize them based on their influence, interest, and potential impact on the project. Tools and methods include:

- Stakeholder Matrix (Influence vs. Interest grid): Helps categorize stakeholders into four quadrants—manage closely, keep satisfied, keep informed, and monitor.
- Venn Diagrams: Visualize relationships and levels of interaction among different stakeholder groups.
- Power-Interest Charts: Determine how much power stakeholders hold and how interested they are in the project outcomes.

This mapping helps in allocating resources efficiently and designing customized engagement approaches for each group.

2. Consultation and Participation

Involving stakeholders through participatory tools ensures transparency, trust-building, and locally-relevant solutions. Effective methods include:

- **Participatory Rural Appraisals (PRAs)**: Allow local people to share, analyze, and plan their development through visual and interactive techniques.
- **Focus Group Discussions (FGDs)**: Provide in-depth insights on issues affecting specific community segments (e.g., women, youth, farmers).
- **Town Hall Meetings and Community Dialogues**: Foster two-way communication and collective decision-making.

Participation should not be symbolic—it must influence project design and decision-making.

3. Capacity Building

Stakeholders must **understand the project and their roles** to participate meaningfully. Capacity-building efforts may include:

- **Workshops and training programs** for farmers, community leaders, extension agents, and local organizations.
- **Orientation sessions** on project goals, rights, responsibilities, and benefits.
- **Technical education** on new agricultural practices, climate-smart technologies, or financial literacy.

When stakeholders are empowered with knowledge and skills, they can co-own the project and drive sustainability.

4. Feedback Mechanisms

Two-way communication is essential. Setting up feedback loops allows stakeholders to voice concerns, suggest improvements, and resolve grievances. Tools may include:

- **Grievance Redress Mechanisms (GRMs)**: Formal systems to handle complaints promptly and fairly.
- **Suggestion Boxes and Hotlines**: Low-barrier platforms for ongoing input.
- **Periodic Review Meetings and Surveys**: Assess satisfaction and engagement levels, and realign strategies as needed.

Transparent handling of feedback fosters trust and accountability.

5. Monitoring Involvement

Stakeholder engagement is not a "set-it-and-forget-it" activity. It must be monitored, evaluated, and adjusted over time. Key steps include:

- Tracking attendance and participation in events and meetings.
- Evaluating engagement quality, not just quantity—were marginalized voices heard?
- Using indicators to measure stakeholder satisfaction, conflict levels, and community support.

Documenting stakeholder involvement helps identify gaps and make informed adjustments throughout the project lifecycle.

2.7 Risks from Poor Stakeholder Engagement

Ignoring or inadequately engaging stakeholders can seriously undermine an agricultural project's success. Poor engagement results in a lack of buy-in, reduced participation, and even active opposition.

Common Risks Include:

• **Resistance or Conflict**

- Stakeholders who feel excluded may oppose the project outright.
- Misinformation, rumors, or lack of clarity can fuel mistrust.

- Example: A community resists land acquisition for an irrigation scheme due to lack of consultation and fear of displacement.

- **Project Delays**
  - Failure to secure permits, local cooperation, or institutional support can halt progress.
  - Unresolved disputes or unaddressed concerns can stall implementation.
- **Inequitable Outcomes**
  - Without proactive inclusion, vulnerable groups—especially women, youth, indigenous populations, and persons with disabilities—may be left out of project benefits.
  - This leads to further social disparities and weakens the overall impact.
- **Sustainability Issues**
  - Projects imposed without local ownership often fail post-closure.
  - If communities are not part of the planning and decision-making, they may not feel responsible for maintaining infrastructure, technologies, or services introduced.

2.8 Best Practices for Stakeholder Management

To ensure strong, equitable, and lasting stakeholder involvement, practitioners should adopt a set of best practices that are both proactive and inclusive.

Recommended Practices:

- **Start Engagement Early**
  - Engagement should begin during the conceptual and planning stages, not after decisions are made.
  - Early involvement ensures local needs shape project design and avoids future resistance.
- **Tailor Communication Strategies**

- Use context-appropriate methods (e.g., local languages, visual aids, radio programs).
- Simplify technical terms and customize messaging to fit different stakeholder literacy levels and cultural norms.

• **Ensure Inclusivity**

- Go beyond token participation—actively engage marginalized groups through separate forums or targeted outreach.
- Encourage diverse representation in project committees, consultations, and benefit-sharing plans.

• **Maintain Ongoing Engagement**

- Continue dialogue throughout implementation, monitoring, and closure.
- Regular check-ins, community forums, and participatory reviews keep stakeholders informed and invested.

• **Document Lessons Learned**

- Keep records of what worked and what didn't in stakeholder engagement.
- Use these insights to improve strategies in future projects, contributing to an organization's institutional learning.

2.9 Summary

Stakeholders form the backbone of any agricultural project. Identifying, understanding, and engaging stakeholders effectively is not just a technical necessity—it's a moral and strategic imperative. Projects that succeed in aligning stakeholder interests, expectations, and contributions are more likely to achieve sustainable, impactful results. The next chapter will explore the project life cycle in agriculture, which ties directly to how and when stakeholders are involved.

Chapter 3: Agricultural Project Life Cycle

Benefit:

Provides a structured methodology for project design and implementation. Students can apply lifecycle tools like Gantt charts, PERT, and CPM in project planning assignments. Researchers can explore project management efficiency or evaluate stage-specific risks.

3.1 Introduction

Every agricultural project, whether it's a smallholder irrigation scheme or a large-scale agribusiness venture, follows a systematic sequence of stages known as the project life cycle. Understanding this cycle is essential for effective planning, resource allocation, stakeholder engagement, and long-term sustainability. This chapter outlines the five major phases of the agricultural project life cycle—Initiation, Planning, Execution, Monitoring & Control, and Closure—and examines how they apply in the agricultural context.

3.2 What is a Project Life Cycle?

The project life cycle refers to the series of phases that a project passes through from its start to its completion. These phases provide a structured approach for managing projects and help project teams achieve objectives efficiently. Each phase has distinct goals, processes, deliverables, and challenges.

In agriculture, the project life cycle must also account for seasonality, biological processes, market timing, and environmental risks, which are often more dynamic than in other sectors.

3.3 Phase 1: Project Initiation

The Project Initiation phase marks the birth of a project. It is the stage where the need, problem, or opportunity is identified and broadly conceptualized. The purpose is to justify why a project should exist and to lay the groundwork for future planning. Though this phase is often brief, it is strategically important, as it influences the direction, relevance, and design of the entire project.

This phase focuses on asking fundamental questions:

- *What problem are we trying to solve?*
- *Who is affected?*
- *What change do we want to see?*
- *Is this project worth exploring further?*

Key Activities

1. **Needs Assessment or Problem Identification**
   - Involves identifying and articulating a core issue or opportunity.
   - Techniques include field surveys, farmer interviews, expert consultations, community forums, and desk research.
   - Helps uncover systemic challenges (e.g., low yields, input inaccessibility, or climate risk exposure).
2. **Situational and Baseline Analysis**
   - Evaluates the current conditions and context of the target area (economic, environmental, technological, and social).
   - Establishes a baseline to compare future project outcomes.
   - May involve collecting data on land use, crop production, infrastructure, farmer incomes, and market access.
3. **Identification of Project Goals**
   - Converts the identified problems or opportunities into high-level objectives.
   - These goals should align with local development priorities, national policies, or donor strategies (e.g., food security, poverty reduction, climate resilience).

4. **Stakeholder Consultations**
   - Engages farmers, community leaders, extension officers, cooperatives, local authorities, NGOs, and potential funders.
   - Ensures the project is inclusive, participatory, and demand-driven, not imposed top-down.
   - Helps in understanding stakeholder expectations, constraints, and willingness to participate.
5. **Preliminary Feasibility Assessment**
   - A quick check on whether the proposed idea is technically and financially reasonable.
   - May include rough cost estimates, land availability checks, or discussions with service providers.
   - Determines if the idea warrants a full feasibility study in the next phase.
6. **Preparation of a Project Concept Note or Proposal**
   - A concise document (typically 2–5 pages) summarizing the rationale, goals, beneficiaries, location, proposed interventions, and estimated cost.
   - Used to present the idea to donors, government agencies, or internal decision-makers for support and approval.

Typical Outputs

- **Project Concept Note**: Describes the project idea, its justification, target beneficiaries, initial goals, and potential interventions.
- **Problem and Objective Trees**: Visual tools that map out the root causes of the problem and how they relate to possible solutions.
  - *Problem Tree*: Shows cause-effect relationships.
  - *Objective Tree*: Translates problems into positive changes or outcomes.

- **Initial Budget Estimate**: Rough cost outline, including infrastructure, equipment, training, and management expenses.
- **Stakeholder Engagement Summary**: Documentation of who was consulted, key concerns raised, and levels of interest or support.

Agricultural Example

A **local NGO** working in a peri-urban farming community conducts a needs assessment and finds that:

- Up to 30% of vegetable crops are lost post-harvest due to lack of proper storage and transport facilities.
- Farmers use traditional storage methods, which are ineffective in hot climates.
- Market access is limited, and incomes are affected.

During stakeholder consultations, farmers express a strong interest in improved storage options and are willing to contribute labor or land for a shared facility.

Based on this, the NGO prepares a project concept note proposing the establishment of a small solar-powered cold storage facility to:

- Reduce post-harvest losses.
- Improve produce shelf-life and quality.
- Strengthen market linkages by enabling bulk sales at better prices.

They also create a problem tree showing how poor storage leads to financial losses and food waste, and an objective tree proposing cold storage as a means to improve livelihoods and food security.

Key Considerations in the Initiation Phase

- **Alignment with policy**: Ensure the project idea aligns with government agricultural policies or donor thematic areas.
- **Local relevance**: The problem and proposed solution should emerge from local needs, not external assumptions.

- **Sustainability from the start**: Think early about how the project will continue beyond external funding.
- **Flexibility**: Keep the concept broad enough to allow for refinement during the planning phase.

3.4 Phase 2: Project Planning

The Project Planning phase is the foundation upon which the success of the entire agricultural project rests. It transforms a conceptual idea into a practical and executable blueprint. A well-structured planning phase ensures that the project is technically sound, economically viable, environmentally sustainable, and socially acceptable.

Key Activities

1. **Full Feasibility Study**

   A comprehensive assessment is conducted across multiple dimensions:

   - **Technical feasibility**: Assesses whether the project can be practically implemented using available technology, labor, and infrastructure.
   - **Financial feasibility**: Evaluates capital requirements, projected costs, expected returns, break-even point, and profitability indicators like NPV, IRR, and payback period.
   - **Market feasibility**: Analyzes supply-demand dynamics, competitor landscape, pricing strategies, and potential buyers.
   - **Environmental and social feasibility**: Identifies potential impacts on ecosystems, water sources, and communities, and outlines mitigation strategies in line with EIA requirements.

2. **Defining Project Scope, Activities, and Deliverables**

   Clearly articulating:

- What the project will and will not cover (scope).
- The sequence of tasks or activities.
- Tangible outputs and outcomes expected at each stage.

3. **Creating a Work Breakdown Structure (WBS)**

   A hierarchical decomposition of the entire project into manageable units:
   - Phases → Tasks → Subtasks.
   - This helps in assigning responsibilities, estimating time and costs, and tracking progress.
4. **Risk Analysis and Mitigation Planning**
   - Identifying internal and external risks (e.g., droughts, input price volatility, policy changes).
   - Categorizing risks based on likelihood and impact.
   - Developing mitigation plans such as crop insurance, diversified income streams, or buffer stocks.
5. **Resource Allocation**
   - Determining the availability and distribution of land, labor, capital, equipment, and inputs (seeds, fertilizers, water).
   - Planning for financing through loans, grants, cooperatives, or private investors.
6. **Timeline and Milestone Planning**
   - Creating visual tools like Gantt charts to plot the project schedule and dependencies.
   - Applying Critical Path Method (CPM) or Program Evaluation and Review Technique (PERT) for analyzing task sequences, durations, and potential delays.
7. **Defining Monitoring and Evaluation (M&E) Indicators**
   - Selecting KPIs (Key Performance Indicators) to track effectiveness, efficiency, and impact.
   - Setting up baseline data and benchmarks for future comparison.

Typical Outputs

- **Detailed Project Proposal or Business Plan**: A comprehensive document outlining goals, strategies, expected outcomes, and financial forecasts.
- **Feasibility Study Report**: Contains technical specifications, financial projections, and sustainability assessments.
- **Logical Framework Matrix (LogFrame)**: Summarizes objectives, indicators, means of verification, and assumptions.
- **Implementation Schedule**: Defines timelines, deadlines, and responsible personnel.
- **Procurement and Resource Plans**: Lists input requirements, sourcing strategy, and cost estimates.

Agricultural Example

A farmers' cooperative proposes to build a cold storage facility to reduce post-harvest losses of perishable crops. During this phase:

- Soil and land suitability tests are conducted for the construction site.
- A market analysis reveals strong demand for cold storage among nearby tomato and mango producers.
- The business plan is developed with capital costs, operating expenses, projected revenue from leasing storage space, and a repayment strategy for bank financing.
- Risk management includes solar energy backup to mitigate power outages and a maintenance schedule to reduce breakdowns.

3.5 Phase 3: Project Execution (Implementation)

Once planning is complete and resources are secured, the project enters the Execution phase, where ideas become reality. This phase involves the actual deployment of human, financial, and physical resources to carry out the planned activities. It is often the most visible and resource-intensive phase, involving day-to-day operations, construction, delivery of inputs, and stakeholder engagement.

Key Activities

1. **Mobilizing Resources and Personnel**
   - Hiring or assigning qualified personnel (e.g., agronomists, technicians, field workers).
   - Mobilizing capital, tools, materials, and equipment.
2. **Construction and Infrastructure Setup**
   - Building essential infrastructure such as irrigation systems, greenhouses, storage units, processing facilities, or farm access roads.
3. **Field Operations**
   - Carrying out core agricultural activities like land preparation, planting, fertilizing, pest control, irrigation, and harvesting.
   - Establishing demonstration plots or pilot farms if the project includes technology testing.
4. **Disbursing Funds and Managing Procurement**
   - Ensuring timely procurement of inputs and services.
   - Managing financial disbursements according to the project's budget and cash flow plan.
   - Complying with procurement regulations (e.g., competitive bidding, transparency, supplier audits).
5. **Managing Human Resources and Technical Operations**
   - Supervising labor, assigning roles, maintaining records of performance.
   - Ensuring health and safety protocols, especially in construction or chemical handling.

Typical Outputs

- **Completed Infrastructure**: Buildings, irrigation systems, processing units, fencing, etc.
- **Delivered Inputs and Services**: Seeds, fertilizers, pesticides, extension services, training sessions.
- **Beneficiary Participation and Training Records**: Attendance sheets, feedback forms, training evaluations.
- **Interim Performance Reports**: Progress updates, challenges encountered, mitigation actions, and financial reports submitted to donors or stakeholders.

Agricultural Example

In the cold storage project example:

- Construction of the storage facility is completed, including insulated walls, cooling systems, and solar power units.
- Staff are trained in the operation of refrigeration equipment, safety procedures, and inventory management.
- Local farmers are enrolled in a membership scheme and educated on the benefits and rules of using the facility.
- Subsidized rates are offered to promote early adoption.
- Regular reports are prepared to track usage rates, feedback from users, and energy consumption trends.

Key Considerations During Execution

- **Quality Control**: Ensure that infrastructure and inputs meet specifications.
- **Cost Monitoring**: Prevent overruns through regular budget reviews.
- **Communication and Coordination**: Maintain strong communication among stakeholders to address issues quickly.

- **Adaptive Management**: Flexibly adjust activities in response to unforeseen challenges (e.g., weather events, supply chain disruptions).

3.6 Phase 4: Monitoring and Control

This phase occurs concurrently with execution and ensures that the project remains on track in terms of time, cost, and quality.

Key Activities:

- Collecting and analyzing performance data
- Comparing actual vs. planned progress
- Revising strategies as needed (adaptive management)
- Engaging stakeholders for feedback
- Managing risks and unexpected events

Tools and Techniques:

- Key Performance Indicators (KPIs)
- Field reports, supervision visits
- Progress dashboards and M&E software
- Mid-term reviews or audits

Typical Outputs:

- Monitoring reports
- Risk mitigation updates
- Financial tracking sheets
- Updated work plans

Agricultural Example:

Monitoring reveals that energy costs for the cold storage are higher than projected. A switch to more efficient solar battery systems is made, saving 20% in operational costs.

3.7 Phase 5: Project Closure and Evaluation

The closure phase involves finalizing project activities, conducting evaluations, and ensuring sustainability or handover.

Key Activities:

- Final performance evaluation (outcomes vs. objectives)
- Financial reconciliation and audit
- Documentation of lessons learned and best practices
- Stakeholder debriefings and satisfaction surveys
- Exit strategy or handover to local stakeholders

Typical Outputs:

- Final evaluation report
- Financial closure report
- Impact assessment
- Project legacy documentation (manuals, training guides)

Agricultural Example:

The cold storage project is handed over to a local cooperative. An evaluation shows a 25% increase in net farmer income and a 35% reduction in post-harvest losses.

3.8 Integration with Agricultural Seasons and Cycles

One unique aspect of agricultural projects is that they must be aligned with biological and climatic cycles, such as:

- Planting and harvesting seasons
- Rainfall patterns and irrigation schedules
- Breeding cycles in livestock
- Market timing for perishable produce

Project planning must therefore be calendar-sensitive, with contingency plans for weather-related disruptions.

3.9 Summary

The agricultural project life cycle provides a structured framework for guiding projects from inception to completion. Each phase—Initiation, Planning, Execution, Monitoring, and Closure—requires specific tools, skills, and stakeholder engagement strategies. Recognizing the dynamic and seasonal nature of agriculture is key to successfully applying the project life cycle in this context. The next chapter will examine the role and structure of feasibility studies, which serve as the foundation for successful planning and investment in agricultural projects.

Chapter 4: Understanding Feasibility Studies

Benefit:

Equips students with a roadmap to perform feasibility studies, a valuable skill in both academia and the field. Researchers can use the framework to assess potential agricultural innovations or interventions.

4.1 Introduction

Before initiating any agricultural project, it is essential to assess whether the proposed venture is realistic, achievable, and sustainable in the long term. This assessment is carried out through a *feasibility study*—a comprehensive and systematic analysis used to evaluate the viability of a project from multiple perspectives. In agricultural project management, such a study is indispensable, as it provides critical insight into the project's potential for success, helping stakeholders avoid costly errors and misguided investments.

Agriculture involves several dynamic and often unpredictable variables such as weather, soil health, pest infestations, and market demand fluctuations. These factors make agricultural projects inherently risky. A feasibility study mitigates these risks by offering a robust decision-making framework grounded in empirical data and strategic evaluation. This chapter delves into the core components, importance, and methodology of conducting feasibility studies, highlighting their integral role in the successful planning and execution of agricultural projects.

4.2 What is a Feasibility Study?

A feasibility study is an analytical tool used to evaluate the potential of a project in a structured, objective, and evidence-based manner. It investigates all essential dimensions of a proposed initiative—technical, financial, economic, legal, environmental, and social—to determine whether the idea is feasible under existing conditions and available resources.

In the agricultural context, feasibility studies are even more vital due to the complexity and variability of biological systems and environmental interactions. For example, a feasibility study for a poultry farm would not only analyze market demand and cost structure but also consider feed availability, disease management systems, waste disposal options, climate conditions, and regulatory constraints. These studies are indispensable for transforming theoretical concepts into practical, scalable, and profitable agricultural ventures.

4.3 Purpose of a Feasibility Study

A well-structured feasibility study performs multiple critical functions, including but not limited to the following:

- **Assess Viability:**

  It determines whether the project is technically possible and financially sound. This includes evaluating if the proposed technology can be implemented efficiently and whether the financial projections indicate profitability and sustainability.

- **Support Decision-Making:**

  By offering data-driven insights, the study aids in making informed choices about whether to proceed with the project as proposed, adjust its scope, or abandon it altogether. It minimizes guesswork and emotional bias in decision-making.

- **Attract Funding:**

  Investors and financial institutions often require a comprehensive feasibility study as a prerequisite for funding. A clear, well-prepared study enhances the project's credibility and showcases its potential return on investment.

- **Identify Risks and Alternatives:**

  The study identifies possible challenges such as market competition, labor shortages, or policy changes. It also proposes mitigation strategies and alternative courses of action to handle unforeseen obstacles.
- **Clarify Scope and Objectives:**

  It defines the project's scale, boundaries, goals, and timelines. A precise scope ensures that resources are allocated effectively and stakeholders remain aligned throughout the project lifecycle.

4.4 Key Questions a Feasibility Study Should Answer

To comprehensively evaluate an agricultural project's feasibility, the study must answer a series of fundamental questions:

1. **Is there a market for the product or service?**
   - Who are the target customers?
   - What is the demand-supply scenario?
   - How competitive is the market environment?
   - Are there seasonal or regional market trends?
2. **Is the technology and infrastructure available and appropriate?**
   - Can existing technology support the proposed operations?
   - Is the necessary infrastructure (roads, storage, irrigation) in place or attainable?
   - How adaptable is the technology to local conditions?
3. **Are the required resources (land, labor, capital) accessible?**
   - Is suitable land available in terms of location, size, and fertility?
   - Is there access to a skilled and affordable labor force?
   - Are the financial resources required for startup and operations obtainable?
4. **Is the project financially profitable?**

- What are the expected costs, revenues, and cash flows?
- What is the break-even point, and how long will it take to reach it?
- Does the return on investment justify the risk?

5. **Will it generate economic and social benefits?**
   - Will the project create jobs or support local businesses?
   - Can it enhance food security, rural development, or income diversification?
   - What is its contribution to the broader agricultural economy?
6. **Is it legally and environmentally compliant?**
   - Does the project meet local, national, and international legal standards?
   - Are there environmental regulations related to water use, pesticide application, or waste management?
   - Have environmental impact assessments been conducted?
7. **Are there risks that could threaten its success, and can they be mitigated?**
   - What are the internal and external risks (e.g., pests, political instability, market crashes)?
   - Are contingency plans in place?
   - What risk management tools (e.g., insurance, diversification) can be employed?

4.5 Components of a Good Feasibility Study

A robust and comprehensive feasibility study is the cornerstone of sound agricultural project planning. It systematically analyzes every dimension of a proposed venture to ensure that all aspects—technical, financial, environmental, legal, and social—are thoroughly considered before any investments are made.

Below are the essential components that should be included in a well-prepared agricultural feasibility study:

4.5.1 Executive Summary

The executive summary provides a succinct overview of the entire feasibility study. Though it appears first, it is usually written last, after all analysis has been completed.

Key elements include:

- A brief description of the proposed project (e.g., dairy farm, greenhouse vegetable production, agro-processing unit).
- Summary of key findings from technical, market, financial, and environmental analyses.
- Assessment of overall viability—whether the project is feasible, conditionally viable, or not recommended.
- Principal recommendations or strategic options for implementation.

Example:

A proposed 50-hectare mango plantation in eastern Kenya shows strong market potential in regional and export markets. Financial projections indicate profitability within four years, with a BCR of 1.8. Technical, environmental, and social factors are favorable. Recommended for implementation with minor adjustments to irrigation design.

4.5.2 Project Description

This section outlines the fundamental characteristics and operational parameters of the proposed project.

Content includes:

- Objectives and goals of the project.
- Location and geographical justification (including maps if necessary).
- Scale and size of operations.
- Duration and key phases of the project lifecycle (e.g., planning, establishment, maturity).
- Expected outcomes and impact.

Also describe:

- The nature of the production system (e.g., organic farming, intensive livestock, hydroponics).
- Value chain activities if agribusiness components (e.g., processing, packaging, export) are involved.

4.5.3 Market Feasibility

Market feasibility assesses the commercial viability of the project by evaluating whether there is a sustainable demand for its outputs and whether the enterprise can compete effectively.

Key elements:

- **Demand and Supply Analysis:** Local, regional, and international demand trends; seasonal demand shifts; current and projected supply.
- **Consumer Preferences:** Taste, quality expectations, and packaging requirements.
- **Market Entry Strategy:** Marketing mix (product, price, place, promotion), branding, certifications.
- **Distribution Channels:** Wholesalers, retailers, export agents, online platforms.

- **Competitive Landscape:** Key competitors, price benchmarking, value-added opportunities.

Tools used: SWOT analysis, Porter's Five Forces, market surveys.

4.5.4 Technical Feasibility

This section evaluates whether the project can be successfully implemented using available technology and resources.

Includes assessment of:

- **Natural Resources:** Land availability, soil health (pH, fertility, drainage), topography, and climatic suitability (rainfall, temperature, wind).
- **Water and Irrigation:** Availability, quality, and efficiency of irrigation systems (e.g., drip, sprinkler).
- **Technology and Infrastructure:** Suitability and availability of equipment, storage facilities, transportation, and energy sources.
- **Production Plan:** Details such as cropping calendar, livestock breeding cycle, feed management plans, or input schedules.

Example:

A drip irrigation system is recommended for the tomato farm to optimize water use during dry seasons. Greenhouse technology may be introduced to extend growing periods.

4.5.5 Financial Feasibility

This component examines whether the project will be financially profitable and sustainable over its life cycle.

Should include:

- **Cost Estimates:** Capital investments (land, machinery, buildings) and operational costs (seeds, feed, labor, utilities).
- **Revenue Projections:** Based on realistic yield, market prices, and sales volume.
- **Break-even Analysis:** Determines when revenues will start to exceed costs.
- **Profitability Indicators:**
    - **Net Present Value (NPV):** Present value of future net cash flows.
    - **Internal Rate of Return (IRR):** The discount rate at which NPV becomes zero.
    - **Payback Period:** Time required to recoup the investment.
    - **Benefit-Cost Ratio (BCR):** Ratio of total benefits to total costs.
- **Financing Plan:** Funding sources, loan terms, interest rates, and repayment schedules.

Note: Sensitivity analysis should be conducted on key financial assumptions such as price and yield variability.

4.5.6 Economic and Social Feasibility

This section assesses broader development impacts beyond profitability.

Key considerations:

- **Employment Creation:** Number and type of jobs created (seasonal, permanent).
- **Income Generation:** For project beneficiaries, including out-growers or contract farmers.
- **Social Inclusion:** Opportunities for women, youth, and marginalized groups.
- **Poverty Alleviation:** Will the project help improve livelihoods and reduce vulnerability?

- **Community Engagement:** Acceptance of the project by local populations and integration with existing community priorities.

Example:

A cassava processing plant is projected to support over 1,000 smallholder farmers and create 75 direct jobs in a rural district, contributing significantly to food security and income diversification.

4.5.7 Environmental and Legal Feasibility

Environmental and legal considerations are especially crucial in agriculture due to potential land degradation, water use conflicts, and biodiversity impacts.

Main elements:

- **Environmental Impact Assessment (EIA):** Required by law in many jurisdictions to assess potential harm to soil, water, flora, and fauna.
- **Waste Management:** Plans for managing organic waste, agrochemicals, and wastewater.
- **Sustainability Measures:** Soil conservation, agroforestry, renewable energy use.
- **Legal Compliance:**
  - Land ownership or leasing status.
  - Water use permits.
  - Agricultural and environmental regulations (e.g., pesticide use laws).
  - Labor laws, health and safety compliance.

4.5.8 Risk Assessment and Mitigation

Every agricultural project faces uncertainties. This component identifies and evaluates potential risks and offers mitigation strategies.

Risk categories:

- **Internal Risks:** Poor management, pest outbreaks, equipment failure.
- **External Risks:** Climate change, price volatility, political instability, trade restrictions.

Tools and techniques:

- **Sensitivity Analysis:** Simulates changes in key variables (e.g., what if prices drop by 20%?).
- **Risk Management Plans:**
  - Crop/livestock insurance.
  - Diversification (e.g., multi-crop systems).
  - Strategic partnerships with buyers or financial institutions.
  - Early warning systems for pests and diseases.

4.5.9 Conclusions and Recommendations

The final section summarizes the study's findings and offers strategic recommendations for stakeholders.

Contents include:

- A clear statement on whether the project is feasible.
- Identification of critical success factors and potential deal-breakers.
- Suggested improvements or modifications to enhance viability.
- Recommendation: proceed, revise, or shelve the project.

Example conclusion:

The proposed organic vegetable farm is financially and technically viable with strong local market demand. However, it is recommended that initial scale be reduced by 20% to manage capital risks and allow for a gradual learning curve.

4.6 Characteristics of a High-Quality Feasibility Study

A high-quality feasibility study is crucial for ensuring that an agricultural project has a solid foundation before substantial investments are made. It provides decision-makers with a thorough understanding of the project's potential, risks, and requirements. Here are the key characteristics that make a feasibility study robust and reliable:

Comprehensive

A good feasibility study addresses all the critical aspects that could affect the success or failure of the project. These include:

- **Technical aspects:** Ensuring that the project can be implemented using the best-suited agricultural practices and technology, considering factors such as soil health, water availability, and crop management techniques.
- **Market analysis:** Evaluating demand and supply dynamics, competitor landscapes, pricing trends, and consumer preferences.
- **Financial projections:** Analyzing capital and operational costs, revenue expectations, profitability, and sustainability.
- **Environmental considerations:** Assessing the environmental impact, sustainability practices, and compliance with legal regulations (e.g., EIA).
- **Social and legal factors:** Understanding the project's contribution to local communities, social equity, and compliance with labor laws, land ownership regulations, and other legal frameworks.

A comprehensive feasibility study ensures no important factors are overlooked, reducing the likelihood of costly surprises down the line.

Objective

A high-quality feasibility study is based on factual data, sound analysis, and realistic assumptions. It must avoid bias and subjective opinions, ensuring that all conclusions and recommendations are drawn from verifiable facts and transparent methodologies.

- **Realistic assumptions** about market growth, crop yields, and operational costs must be based on reliable sources such as historical data, scientific research, and expert opinions.
- **Objective criteria** for evaluating success should include measurable and quantifiable indicators such as return on investment (ROI), break-even points, and cash flow projections.

This objective approach fosters credibility and helps stakeholders make informed, evidence-based decisions.

Tailored

Each agricultural project is unique, and its feasibility study should be tailored specifically to its context. A generic feasibility study is not suitable for a specific project because agricultural ventures are affected by local conditions such as climate, soil types, market access, and community dynamics.

- **Site-specific information** is critical for the study. For example, a feasibility study for a fruit farm in a tropical climate will differ significantly from one for a grain farm in a temperate zone.
- **Cultural and local knowledge** should be integrated into the study, as understanding local practices, resources, and priorities can determine the success of the project.
- **Adaptation to local policies** and regulations, such as land tenure systems, water use laws, and labor laws, is essential for understanding potential challenges or obstacles.

Tailoring the feasibility study to the specific project ensures that all relevant factors are considered.

Data-Driven

A well-supported feasibility study is backed by credible, up-to-date data. This includes field surveys, consultations with experts, market research, and secondary data sources such as government reports, agricultural studies, and industry analyses.

- **Primary data** might come from surveys with farmers, interviews with industry experts, or direct observations of production practices.
- **Secondary data** can include published research, industry reports, government statistics, and academic studies.
- **Verification and validation** of data ensure that assumptions are based on real-world information, reducing the risk of faulty conclusions.

Data-driven feasibility studies are grounded in reality, providing stakeholders with confidence in the analysis and recommendations.

Actionable

The feasibility study should not just present facts but also offer practical recommendations and strategies for project implementation. It should outline actionable steps and provide clear guidance on what needs to be done next.

- **Clear decision-making paths** should be laid out. For example, if the study finds that the project is not financially viable, it should provide alternatives or suggestions for revising the business model.
- **Implementation guidelines** should highlight key steps, timelines, and responsibilities, including a breakdown of tasks such as land preparation, securing financing, obtaining permits, or sourcing materials.
- **Risk management** strategies should be offered to mitigate potential challenges, such as weather variability or market price fluctuations.

By providing actionable recommendations, the feasibility study helps project managers and investors confidently proceed with implementation or make adjustments if necessary.

4.7 Common Mistakes in Feasibility Studies

While feasibility studies are invaluable tools for project planning, they are only effective if done correctly. Several common mistakes can undermine the reliability of a feasibility study:

1. **Relying on Outdated or Overly Optimistic Data**

   Using outdated data or making overly optimistic assumptions about market conditions, yields, or financial returns can lead to inaccurate conclusions. It's essential to use the most current data and adopt conservative assumptions to avoid overestimating potential returns.

2. **Ignoring Local Realities**

   Failing to account for local cultural, environmental, or logistical factors can cause projects to falter. For example, understanding local farming practices, community preferences, and regional climate conditions is crucial for designing a project that aligns with the area's needs and capabilities.

3. **Underestimating Costs or Overestimating Revenues**

   Accurately estimating costs is critical. Underestimating the cost of inputs, labor, or infrastructure can result in budget overruns. Similarly, overestimating potential revenues or market prices can lead to an overly optimistic financial outlook. Both miscalculations can result in the project running into financial difficulties later on.

4. **Failing to Involve Stakeholders**

   Stakeholder engagement should be an integral part of the feasibility study. Local farmers, community leaders, and other stakeholders often have valuable insights that can refine the project and increase its chances of success. Ignoring their input may result in missed opportunities or community resistance.

5. **Skipping Risk Analysis or Assuming "Business as Usual"**

   Every project faces risks, from climate change to market fluctuations. Failing to identify and plan for potential risks can leave a project vulnerable to unforeseen challenges. Risk analysis is essential to ensure that the feasibility study accounts for uncertainties and offers mitigation strategies.

Remember: A flawed feasibility study is more harmful than none at all. It can lead to costly failures, loss of investor confidence, and wasted resources.

4.8 Summary

A feasibility study is a critical tool for assessing the viability and potential success of an agricultural project. It provides stakeholders with an in-depth, fact-based analysis of the project's technical, financial, market, social, and environmental feasibility. By offering clear, actionable recommendations and identifying potential risks, a feasibility study helps ensure that resources are invested wisely and that projects are implemented with a clear understanding of their challenges and opportunities.

The next chapters of this book will provide a detailed examination of each component of the feasibility study, starting with Market Feasibility Analysis in Chapter 5. By delving deeper into specific areas of study, readers will gain a comprehensive understanding of the steps required to develop a successful agricultural project.

Part II: Feasibility Study and Analysis

Focus: In-depth exploration of feasibility dimensions and evaluation techniques

Chapter 5: Market Feasibility Analysis

Benefit:

Introduces students to market research methods, such as demand analysis and competitor profiling. Ideal for those conducting agribusiness studies or analyzing value chains. Researchers can incorporate these tools into value chain development or rural marketing strategies.

5.1 Introduction

In agricultural projects, having a technically sound approach is not enough to guarantee success. Even if the production process is efficient, the project may fail if there is no viable market for its output. This is where market feasibility analysis becomes crucial. A comprehensive market feasibility analysis provides insights into the demand for the agricultural products, the competitive environment, and the logistical challenges involved in delivering the product to the target customers.

In this chapter, we will explore the critical components of a market feasibility analysis, methodologies for conducting the analysis, and its application in agricultural project planning. By understanding the market dynamics, agricultural entrepreneurs, farmers, and investors can better evaluate the risks and opportunities associated with the commercialization of agricultural outputs.

5.2 What is Market Feasibility?

Market feasibility refers to the process of evaluating the commercial viability of an agricultural product or service in a specific market. It assesses whether there is sufficient demand for the product, if the market can support the proposed scale of operations, and if the product can be sold at a profitable price.

Market feasibility analysis takes into account factors such as market size, demand trends, competition, consumer behavior, and distribution channels. In agriculture, market feasibility is especially vital because agricultural products are subject to seasonal variations, fluctuating market prices, and changing consumer preferences. Therefore, the success of an agricultural project is often determined by the project's ability to meet market demand effectively.

A well-conducted market feasibility study ensures that the project is not only technically viable but also commercially sustainable. It helps stakeholders understand whether the investment is likely to result in profitability or if the project should be adjusted or even abandoned.

5.3 Key Objectives of Market Feasibility Analysis

The goal of a market feasibility analysis is to answer critical questions about the commercial viability of the proposed agricultural project. Below are the key objectives that such an analysis seeks to fulfil:

Determine Who the Customers Are (Local, Regional, International)

Understanding the target customers is the first step in assessing market feasibility. In agricultural projects, customers can be categorized into local, regional, and international markets, each with its own dynamics.

- **Local Markets**: These could include local consumers, small retailers, farmers' markets, and community-based supply chains. The advantage of local markets is lower transportation costs and the ability to build relationships directly with consumers.
- **Regional Markets**: Often more competitive than local markets, regional markets include larger towns, cities, and regions where demand for agricultural products may be higher due to population density and greater purchasing power.

- **International Markets**: If the project is considering export, the international market becomes important. Understanding global trade regulations, export standards, and foreign demand is critical for success in international markets.

A key task in market feasibility analysis is identifying which segment of the market offers the best potential for the agricultural project, based on factors like product demand, pricing, and access to distribution channels.

Understand What Customers Want, How Much, and at What Price

Once the target market is identified, the next step is to understand customer preferences, including product specifications (e.g., organic versus conventional), quantities required, and price points that are acceptable.

- **Product specifications**: What are the specific requirements that consumers have? For example, consumers might demand high-quality, pesticide-free produce, or they may prefer products with specific certifications like organic or fair trade.
- **Demand volume**: Estimating the volume of product that customers are likely to purchase in a given timeframe is essential for understanding the scale of production needed. It also helps with inventory planning, resource allocation, and managing supply chain logistics.
- **Pricing**: Analyzing the acceptable price range for the product is crucial. Setting a price that is too high may alienate potential customers, while setting it too low could lead to unsustainable profits. A competitive pricing strategy should consider production costs, competitor prices, and market conditions.

Customer research—such as surveys, interviews, and focus groups—is a valuable tool in gathering this information to make informed decisions.

Analyze Existing Competitors and Their Market Share

Understanding the competitive landscape is vital for any agricultural project. The market may already be saturated with similar products, or there may be gaps in the market that present opportunities. A thorough competitor analysis involves:

- **Identifying key competitors**: Who are the main players in the market? What products do they offer, and what is their market share?
- **Evaluating competitor strengths and weaknesses**: What are the competitors doing well, and where are they falling short? This can help identify opportunities for differentiation and potential competitive advantages.
- **Pricing strategies**: How do competitors price their products? Are they using aggressive pricing tactics, or do they focus on premium pricing? Understanding competitor pricing strategies helps ensure that the new project can position itself effectively in the market.

The insights gained from a competitor analysis can inform decisions about how to differentiate the product and establish a competitive edge.

Identify Market Gaps and Opportunities for Differentiation

In competitive markets, finding a unique selling proposition (USP) is key to success. Market feasibility analysis should identify potential gaps in the market and areas where the agricultural product can stand out.

- **Market gaps**: These could be in underserved regions, customer segments, or specific product offerings that competitors have overlooked. For example, there might be a growing demand for gluten-free products or specific varieties of organic produce that are not being met by existing suppliers.
- **Opportunities for differentiation**: Differentiation can be achieved through factors such as product quality, branding, packaging, sustainable practices, or customer service. For example, a farm could offer a premium

product by using organic farming methods or implement innovative packaging to appeal to eco-conscious consumers.

By identifying gaps and opportunities for differentiation, the project can carve out a niche in the market and attract a loyal customer base.

Evaluate the Distribution and Logistics System for Getting Products to Market

Market feasibility is not only about demand but also about the logistics of getting the product from farm to market. A comprehensive assessment of distribution channels is essential for understanding the costs and logistics involved in delivering the product to customers.

- **Local distribution**: For local markets, the focus may be on short supply chains involving farmers' markets, small retailers, or direct-to-consumer sales.
- **Long-distance distribution**: For regional or international markets, logistics becomes more complex and may involve third-party distributors, wholesalers, and cold chain systems.
- **Distribution costs and infrastructure**: Understanding the cost of transportation, warehousing, and packaging is crucial for determining the feasibility of reaching customers at a profit.

An effective logistics strategy ensures that products arrive on time, in good condition, and at the lowest possible cost.

Project Potential Sales Volume and Revenue

One of the most important objectives of market feasibility is estimating potential sales volume and revenue. This involves forecasting:

- **Sales volume**: Based on market size, customer demand, and competitor activity, the feasibility study should estimate how much product can realistically be sold over a given period (e.g., annually or seasonally).

- **Revenue projections**: Once sales volume is estimated, revenue projections can be made by multiplying the expected sales volume by the anticipated price point.

These projections help establish financial expectations, inform investment decisions, and guide cash flow management.

5.4 Components of Market Feasibility Analysis

A thorough market feasibility analysis involves several key components that collectively assess the likelihood of success in the marketplace. These components help ensure that the agricultural project aligns with demand, competition, pricing, customer preferences, and distribution channels. Each section provides critical insights into how the market will absorb the agricultural product and whether the project is commercially viable.

5.4.1 Demand Analysis

Demand analysis is a critical part of the market feasibility process, as it focuses on understanding how much of the agricultural product is needed and by whom. It helps identify whether there is a sustainable market demand for the product at the proposed production scale.

Key Factors to Assess:

- **Historical Consumption Trends**: Reviewing past consumption patterns helps forecast future demand. This can provide insights into whether the product has been stable in the market or if demand is seasonal or cyclical.
- **Current Demand in Target Market**: Assessing how much of the product is currently being consumed in the target market. This includes identifying whether the product is in short supply or experiencing growing demand.

- **Seasonal Variations**: Agricultural products often have seasonal peaks in demand. For example, certain fruits or vegetables may only be in demand during harvest seasons. Understanding these patterns helps in planning the timing of production and sales.
- **Purchasing Power and Consumer Preferences**: The willingness and ability of the target market to buy at certain price points is crucial. Additionally, preferences (e.g., organic vs. conventional products) can greatly influence demand.
- **Market Segments**: Identifying the type of customers, such as wholesalers, retailers, or end consumers, can help in determining where to focus marketing efforts.

Tools:

- **Market Surveys**: Surveys of potential customers can gather firsthand data about demand patterns and preferences.
- **Interviews with Buyers and Traders**: Direct discussions with key market players provide qualitative insights.
- **Government Agricultural Data**: Data on past production and consumption can guide demand estimations.
- **Trade Data**: If considering exports, trade data can help evaluate international demand and competition.

Example:

If planning to grow onions, assess whether local demand exceeds current supply, particularly during off-season periods when onions may not be available from other sources. Understanding the demand fluctuations can help determine if there's a market gap that the project can fill.

5.4.2 Supply Analysis

Understanding the current and potential supply situation in the target market helps gauge the level of competition and whether the market is already saturated or has room for additional products.

Key Factors to Assess:

- **Local Production Levels and Major Producers**: Identify the key producers in the area and their output levels. A high concentration of producers might indicate a saturated market, while a low supply could indicate an opportunity.
- **Import Competition**: If there are cheaper or higher-quality imported goods available, they may compete with local production. Understanding the price and availability of imported goods is important for determining your competitive edge.
- **Substitutes**: Assess whether other products can replace the one being produced. For example, sweet potatoes may compete with yams, or locally produced rice may compete with imported varieties.
- **Value Chain Gaps**: Look for inefficiencies or gaps in the local value chain, such as inadequate post-harvest storage or a lack of processing capacity, which may provide opportunities for value addition.

Example:

If tomato production peaks during a specific season, your project could either target an off-season production window or invest in preservation/processing (such as canning or drying) to stabilize supply throughout the year.

5.4.3 Price Analysis

Price analysis helps determine the potential revenue from the agricultural project and whether the selling price will provide sufficient margins to cover costs and generate profit. Understanding price fluctuations and trends is crucial for setting competitive and sustainable prices.

Key Elements to Assess:

- **Historical and Current Market Prices**: Reviewing how prices have fluctuated in the past helps project future pricing trends.
- **Price Volatility and Seasonality**: Prices for agricultural products often vary by season and market conditions. Understanding these price swings allows the project to anticipate low-price periods and plan accordingly.
- **Cost vs. Selling Price**: Analyzing production costs in relation to market prices helps determine the profit margin. If production costs are too high relative to selling prices, the project may be unsustainable.
- **Farmgate vs. Wholesale vs. Retail Price Levels**: Different levels of the value chain (farmgate, wholesale, retail) have different price points. Understanding these can help identify the most profitable channel for sales.

Tools:

- **Price Trend Analysis**: Use graphs or spreadsheets to analyze historical price movements and identify trends.
- **Field Visits to Marketplaces**: On-the-ground visits to local markets or wholesalers provide real-time data on pricing.
- **Interviews with Buyers and Processors**: Engage with buyers, traders, and processors to understand their price expectations.

Example:

If cassava prices drop by 30% during harvest season, exploring value addition (e.g., converting cassava into flour) or delayed processing may help stabilize income during price dips.

5.4.4 Customer Analysis

Understanding the customer profile is essential for aligning the agricultural product with the needs and preferences of the market. Customer analysis helps identify the specific buyers and their purchasing behavior.

Key Considerations:

- **Type of Buyers**: Identify the different types of buyers who will purchase the product, such as wholesalers, processors, retailers, or export firms.
- **Buying Behavior and Quality Preferences**: Assess what factors influence customer decisions—price sensitivity, product quality, packaging, and origin (e.g., preference for organic produce).
- **Certification and Standards**: Many buyers, especially export firms, may require certification (e.g., organic, fair trade) to ensure the product meets specific standards.

Example:

Exporters of mangoes may prioritize organic certification for quality assurance, while local consumers may be more concerned with price and freshness. Understanding these nuances allows the project to tailor its approach to different segments.

5.4.5 Competitive Analysis

Understanding the competition is a fundamental part of market feasibility. By evaluating who else is offering similar products and what their strengths and weaknesses are, a project can find ways to differentiate itself and create a competitive advantage.

Key Elements to Analyze:

- **Number and Size of Competitors**: Identify how many competitors exist in the market and how large they are. A highly competitive market might require differentiation strategies, while a less competitive market could indicate a market gap.
- **Market Share Distribution**: Understanding how the market is divided among competitors helps gauge the level of market control each player has.

- **Unique Selling Points (USPs)**: Analyze what makes your product different from others in the market. It could be related to product quality, certifications, pricing, or packaging.
- **Competitor SWOT Analysis**: Evaluate competitors' strengths, weaknesses, opportunities, and threats (SWOT analysis) to better understand how your project can offer something better or different.

Example:

In a saturated market of white maize producers, shifting focus to a specialty product, such as high-protein maize or a drought-resistant variety, could provide a niche opportunity with less direct competition.

5.4.6 Distribution and Marketing Channels

The feasibility of distributing agricultural products and getting them to customers in a timely and cost-effective manner is a crucial component of market feasibility. A strong distribution and marketing strategy ensures that the product reaches the right consumers at the right time.

Key Factors to Assess:

- **Transport Infrastructure and Costs**: Evaluate the availability and quality of transportation networks. Poor infrastructure can increase costs and delays, impacting the profitability of the project.
- **Cold Chain or Storage Needs**: For perishable goods, assess whether cold storage or refrigeration is required to preserve product quality.
- **Middlemen Involvement and Markup**: Understand the role of intermediaries (e.g., wholesalers, traders) and how their involvement may affect the price structure and profitability.
- **Direct vs. Indirect Sales Models**: Consider whether to sell directly to consumers, through cooperatives, or via contract farming agreements. Each model has different implications for control, profit margins, and logistics.

Example:

For perishable fruits like berries, having access to refrigerated transport or urban farmers' markets is essential to maintain product freshness and reduce spoilage.

5.5 Tools and Methods for Market Analysis

Tool	Use
Surveys and Questionnaires	Directly collect customer and trader feedback
Focus Group Discussions	Understand preferences and price sensitivity
Government Market Reports	Provide reliable secondary data
Value Chain Mapping	Visualize flows and key players in the market
SWOT and PEST Analysis	Assess internal/external market factors
Field Observation	Identify informal trends, quality standards, and price negotiation behaviors

Market analysis involves gathering and interpreting data to make informed decisions about the commercial viability of an agricultural project. To gain accurate insights, various tools and methods can be employed. Each of these tools helps focus on different aspects of the market, from consumer preferences to competitive forces. Here's an in-depth look at some of the key tools and methods used for market analysis:

Surveys and Questionnaires

Purpose: Surveys and questionnaires are primary tools for collecting direct feedback from potential customers, traders, and other market participants.

Use:

- **Customer Feedback**: Surveys and questionnaires allow project planners to directly ask potential consumers about their preferences, willingness to pay, buying habits, and product expectations.

- **Trader and Supplier Insights**: These tools can also be used to gauge the opinions of wholesalers, retailers, and other key intermediaries in the value chain. For example, a survey could inquire about their supply challenges, product quality expectations, or seasonal demand.
- **Quantitative Data**: These tools can be structured to gather measurable data (e.g., average prices, quantities purchased, frequency of purchase) that can be analyzed statistically.

Example:

If planning to introduce a new type of organic tomato, a survey could assess whether local consumers are willing to pay a premium for organic produce, and if so, what price range they consider acceptable.

Focus Group Discussions (FGDs)

Purpose: Focus group discussions involve a small group of individuals from the target market (e.g., customers, farmers, traders) who are invited to discuss specific topics related to the agricultural product.

Use:

- **Preference Exploration**: FGDs are particularly useful for understanding the preferences and attitudes of consumers. They provide insights into what drives purchasing decisions, such as taste, quality, sustainability, or price.
- **Price Sensitivity**: Group discussions can also help gauge how price-sensitive the target audience is. By discussing the pros and cons of different pricing options, participants can reveal their thresholds for acceptable price points.
- **Qualitative Insights**: While surveys provide quantitative data, FGDs generate qualitative insights that are valuable for understanding deeper market dynamics and emotional drivers of consumer behavior.

Example:

A focus group with local farmers could reveal preferences for certain varieties of maize based on factors like resistance to pests, yield, or the ease of harvesting, which could guide product development.

Government Market Reports

Purpose: Government market reports provide secondary data, which can be an invaluable source of information for market analysis.

Use:

- **Reliable Secondary Data**: These reports typically offer reliable data collected from national or regional agricultural surveys, including production statistics, consumption trends, and market forecasts.
- **Market Insights**: Government reports can help identify overarching trends, such as changes in consumer preferences, supply chain challenges, or evolving agricultural policies. They often contain data on pricing trends, market share, and competitor analysis at a macro level.
- **Legal and Regulatory Context**: These reports also include information on regulatory changes that might affect market access, such as new trade tariffs or import restrictions, which are especially relevant for export-oriented projects.

Example:

A government report on the tomato market could reveal a decrease in local production due to a disease outbreak, providing an opportunity for new entrants who can offer disease-resistant varieties.

Value Chain Mapping

Purpose: Value chain mapping visually outlines the entire process of how an agricultural product moves from production to consumption, identifying key players and stages in the process.

Use:

- **Understanding the Flow**: Value chain mapping helps identify how raw materials or products move through various stages, such as production, processing, transportation, and retailing.
- **Identifying Gaps and Opportunities**: By visualizing each stage in the value chain, you can identify inefficiencies or bottlenecks that may offer opportunities for improvement. For example, a lack of post-harvest storage could create opportunities for refrigerated storage solutions.
- **Key Players**: The map can also highlight important stakeholders—producers, traders, processors, wholesalers, and retailers—and illustrate how they are connected.

Example:

A value chain map for a banana export project might reveal that farmers rely on intermediaries to reach international markets, which could prompt the project to consider direct export channels to reduce costs.

SWOT and PEST Analysis

Purpose: SWOT (Strengths, Weaknesses, Opportunities, Threats) and PEST (Political, Economic, Social, Technological) analyses are tools used to assess both internal and external market factors.

Use:

- **SWOT Analysis**: SWOT helps assess the internal strengths and weaknesses of a project or business, as well as external opportunities and threats in the market environment. For example, strengths might include a unique product offering, while threats could be high competition or market volatility.
    - **Strengths**: What advantages do you have? (e.g., experienced management, access to land)

    - **Weaknesses**: What challenges do you face? (e.g., limited access to capital)
    - **Opportunities**: What market opportunities exist? (e.g., increasing demand for organic products)
    - **Threats**: What external factors could harm the business? (e.g., fluctuating market prices, climate change)
- **PEST Analysis**: PEST analysis looks at the macroeconomic environment that could influence market feasibility. This tool evaluates how political, economic, social, and technological factors affect the market.
    - **Political**: What regulations, government policies, or trade agreements affect the industry?
    - **Economic**: What economic trends (e.g., inflation, unemployment) affect market demand?
    - **Social**: How do social trends (e.g., consumer preferences for sustainable products) impact the market?
    - **Technological**: What technological advancements (e.g., new agricultural equipment or processing methods) affect production or market access?

Example:

A SWOT analysis might reveal that while a project has a strong supply of labor (a strength), the reliance on unpredictable rainfall (a weakness) makes the project vulnerable to climate change (a threat). A PEST analysis could indicate that government subsidies for organic farming (an opportunity) could create a favorable environment for the project.

Field Observation

Purpose: Field observation involves gathering real-time insights from markets, farms, or processing centers to identify informal trends, behaviors, and quality standards that may not be captured through surveys or reports.

Use:

- **Understanding Market Behavior**: By observing how products are sold, negotiated, and valued in the marketplace, you can gain insights into factors such as consumer trust, price negotiation, or quality control standards.
- **Identifying Informal Market Trends**: Some trends, such as local demand for specific varieties of crops, may not be documented in formal reports but can be uncovered through direct observation.
- **Quality Standards and Product Perceptions**: Field visits help assess what quality standards exist in the market and how products are perceived by consumers. For example, are there informal grading systems used by traders or consumers?

Example:

A visit to a local farmers' market could reveal that consumers prefer smaller, organic tomatoes, which may not have been reflected in official surveys. This insight could guide production and marketing strategies to meet consumer preferences.

5.6 Projecting Market Potential and Revenue

Once demand and pricing are understood, project developers can estimate:

- **Target market share** (realistic based on production scale)
- Expected volumes sold
- Revenue forecasts (volume × average price)
- Sales growth potential (expansion scenarios)

5.7 Risks and Challenges in Market Feasibility

Risk	Example
Price Fluctuations	Market price for produce crashes during peak harvest
Access to Market	Poor roads or distance from main trading centers
Buyer Dependence	Relying on a single bulk buyer who changes terms
Import Competition	Imported goods undercutting local prices
Changing Consumer Trends	Shifts in taste or dietary habits (e.g., more plant-based diets)

Market feasibility analysis not only identifies opportunities for growth and success but also highlights potential risks that could undermine the success of an agricultural project. Here are some key risks and challenges that can impact the commercial success of agricultural ventures:

Price Fluctuations

Risk: Price fluctuations refer to the variability in the market price of agricultural products due to various factors such as supply and demand changes, seasonality, or external market influences.

Example:

If a crop like tomatoes is grown in abundance during the peak season, the market may experience a glut, causing prices to plummet. Conversely, if there is a shortage due to bad weather or pests, prices may rise significantly. These fluctuations can drastically affect the profitability of the project.

Access to Market

Risk: Limited or difficult access to markets can result from inadequate infrastructure, such as poor roads, insufficient transport options, or distance from urban centers where demand is higher.

Example:

A farmer in a remote rural area may struggle to reach buyers in major cities due to poor road conditions or long transportation times, which can increase costs and reduce the competitiveness of their product.

Buyer Dependence

Risk: Relying on a single buyer, especially a large corporation or wholesaler, can be risky if the buyer changes terms or stops purchasing altogether. This creates a situation where the project is highly vulnerable to the buyer's decisions.

Example:

An agribusiness that supplies fruits to a single supermarket chain may face severe disruptions if the supermarket decides to switch suppliers or renegotiates the purchase terms, leaving the agribusiness without a stable revenue stream.

Import Competition

Risk: Imported goods can often undercut local prices, especially if the imported products are cheaper due to lower production costs abroad or subsidies in the exporting country. Local farmers may struggle to compete with these lower-priced goods.

Example:

Local maize producers may find it difficult to sell their produce at competitive prices if cheaper maize is imported from countries with lower labor and production costs, reducing the domestic market share for locally grown maize.

Changing Consumer Trends

Risk: Changes in consumer preferences, tastes, or dietary habits can impact demand for certain products. If the market shifts away from a specific product, it could leave producers with unsold goods or force them to reorient their business model.

Example:

If there is a sudden increase in demand for plant-based foods, traditional meat producers may face reduced demand, as consumers opt for alternatives like plant-based proteins or dairy substitutes.

5.8 Mitigation Strategies

To address these risks, agricultural project planners can adopt several strategies to reduce market-related uncertainties and increase the likelihood of success:

Establish Forward Contracts with Buyers

Strategy:

Forward contracts allow farmers or producers to secure agreements with buyers at predetermined prices for a specific quantity of produce in the future. This helps mitigate the risk of price fluctuations and ensures market access.

Benefit:

Farmers can lock in prices for their products before they harvest, providing financial stability and reducing exposure to price volatility.

Example:

A farmer growing coffee may sign a forward contract with a coffee roaster to sell a fixed quantity at a guaranteed price, ensuring income stability even if market prices fluctuate.

Use Cooperative Marketing for Better Bargaining Power

Strategy:

By pooling their resources, small-scale farmers can form cooperatives to collectively market their products. This approach allows for better bargaining power, reducing dependence on a single buyer and increasing access to larger markets.

Benefit:

Cooperatives can negotiate better prices, access larger distribution channels, and reduce transportation costs through shared logistics.

Example:

A group of vegetable growers can form a cooperative, allowing them to sell their produce in bulk to supermarkets or export markets, thus achieving better prices than they would individually.

Invest in Value Addition (e.g., Drying, Processing, Branding)

Strategy:

Value addition involves processing raw agricultural products into higher-value goods, such as drying fruits, making sauces, or branding products with unique qualities. This reduces vulnerability to price fluctuations and offers a way to capture more value from the product.

Benefit:

Value-added products often have higher profit margins and are less prone to market price fluctuations because they are differentiated.

Example:

A mango producer can process and package dried mango slices or mango puree, targeting both local and export markets. This product has a longer shelf life and can be sold at a premium price compared to fresh mangoes.

Target Off-Season Production

Strategy:

To address market demand during periods when supply is low, producers can target off-season production or use storage technologies. This allows them to sell their products at higher prices when competition is less.

Benefit:

Off-season products can command premium prices because they fill gaps in the market when other producers are unable to supply.

Example:

A tomato grower can use greenhouse technology to produce tomatoes during the off-season, when local production is low, and command higher prices due to scarcity.

Build Market Intelligence Systems to Track Trends

Strategy:

Implementing a market intelligence system enables producers to monitor changing consumer preferences, competitor strategies, and market conditions in real time. This information helps in adapting quickly to shifts in the market.

Benefit:

With up-to-date market data, producers can adjust their strategies, such as pricing, marketing, and product offerings, to better align with current trends.

Example:

A dairy farmer could use market intelligence systems to track the increasing demand for plant-based milk and, if necessary, diversify into producing almond or oat milk to capture new market segments.

5.9 Summary

Market feasibility analysis is a critical step in determining whether an agricultural project can succeed commercially. It helps answer the key question of whether the product will sell and at a profitable price. By understanding demand, supply, pricing, consumer behavior, and competition, project planners can design strategies to mitigate risks such as price fluctuations, market access issues, and buyer dependence.

The next chapter will focus on Technical Feasibility, where we will explore whether the project is physically and technologically feasible, including factors like land suitability, infrastructure, and available technologies.

Chapter 6: Technical Feasibility

Benefit:

Encourages practical understanding of agricultural inputs and infrastructure needs. Students can evaluate land and water suitability in lab/field studies. Researchers can explore the relationship between tech adoption and productivity.

6.1 Introduction

In any agricultural project, technical feasibility is a crucial factor that determines whether the project can be implemented efficiently, effectively, and sustainably with the available resources. Even if the market potential and financial viability of a project are promising, it will fail without the proper infrastructure, technology, land, and skills needed for success. This chapter dives into the technical aspects that must be considered when evaluating the feasibility of an agricultural venture, ensuring that all technical challenges can be addressed before moving forward with implementation.

6.2 What is Technical Feasibility?

Technical feasibility refers to the evaluation of whether an agricultural project can be executed successfully using the available technology, land, resources, and labor, while operating under the specific environmental and climatic conditions of the region. Essentially, it asks: Can the project be practically carried out with the existing tools and systems? This includes evaluating the ability to use the land effectively, ensuring that adequate technology and inputs are available, and confirming that the workforce possesses the required skills or can be trained to meet project needs.

It is important to assess both the production capacity (how much the project can produce over time) and the operational practicality (whether the project can run efficiently and within its defined scope). A technically feasible project is one that can function smoothly under real-world conditions, ensuring that production targets are met with minimal risk of failure.

6.3 Objectives of Technical Feasibility Analysis

A technical feasibility analysis aims to answer key questions related to the resources required and their availability for successful project implementation. The objectives include:

- **Land, Soil, and Climate Suitability**: The project must be placed in a location with the right soil type, climate, and land conditions to support the chosen crops or livestock.
- **Availability of Key Inputs**: Essential agricultural inputs, such as seeds, fertilizers, pesticides, machinery, and labor, must be available and reliable.
- **Infrastructure and Equipment**: Identify the infrastructure required for the project, such as storage, processing facilities, roads, and equipment. Assess whether these can be sourced locally and at reasonable costs.
- **Human Resources**: Ensure that the necessary technical expertise, such as skilled labor or specialized knowledge, is available. If not, determine if training programs or recruitment strategies are feasible.
- **Production Efficiency and Sustainability**: Evaluate whether the proposed agricultural methods are feasible, efficient, and sustainable in the long term, considering environmental impact, cost-effectiveness, and potential for scaling.

6.4 Components of Technical Feasibility

The following components are essential to evaluating the technical feasibility of an agricultural project:

6.4.1 Land Suitability and Soil Analysis

Land and soil conditions are crucial to the success of any agricultural project, as they directly affect crop yield and quality. A comprehensive assessment of the land and soil is essential to determine whether the proposed agricultural activities are feasible.

Key Factors to Assess:

- **Topography**: The land's slope impacts how easily it can be farmed, particularly regarding mechanization and erosion risk. Flat land is generally preferred for large-scale mechanized farming, while sloped land may require terracing or specialized equipment.
- **Soil Texture and Structure**: The soil's composition, such as whether it is sandy, loamy, or clayey, affects water retention, drainage, and root growth. Sandy soils drain quickly, while clayey soils retain more water, potentially leading to waterlogging if not managed properly.
- **Soil pH**: Soil pH affects nutrient availability to plants. Most crops prefer slightly acidic to neutral soils, but some, like blueberries or coffee, may thrive in more acidic soils.
- **Fertility and Nutrient Levels**: Soil fertility, which can be determined through soil testing, is critical for healthy crop growth. Soil tests typically measure nitrogen (N), phosphorus (P), and potassium (K), along with other micronutrients. Fertility management strategies (such as organic or chemical fertilization) will be necessary if deficiencies are detected.
- **Drainage**: Poor drainage can cause waterlogging, which may lead to root rot or other diseases. Proper land leveling, drainage systems, or raised beds might be necessary in areas with low-lying or poorly draining soils.

Tools for Soil and Land Assessment:

- Soil testing kits or laboratory analysis for pH, fertility, and nutrient levels.
- Land classification maps to understand soil types, slopes, and land suitability.

- Agro-ecological zoning to determine the appropriateness of land for different types of agriculture, based on climate and soil conditions.

Example:

A coffee plantation is best situated in areas with slightly acidic soil, good drainage, and moderate elevation. If the soil has poor drainage or is alkaline, adjustments (like adding sulfur to acidify the soil) or a different location may be necessary.

6.4.2 Water Availability and Irrigation Systems

Water is an essential input for agriculture, especially in regions with varying rainfall patterns. A detailed assessment of water availability and the irrigation system options is necessary to ensure consistent and reliable water supply to the crops or livestock.

Key Factors to Assess:

- **Sources of Water**: Water can come from various sources, such as rivers, boreholes, reservoirs, or rainwater harvesting systems. It's essential to assess whether these sources are reliable and sufficient for the scale of the project.
- **Quantity and Timing**: Seasonal fluctuations, drought risks, and the amount of water available during peak growing periods must be considered. For instance, crops may require more water during the dry season, making access to adequate water supply crucial.
- **Water Quality**: Assessing water quality is vital, as high salinity or contamination can damage crops or livestock. Testing water for contamination and salinity levels ensures it is suitable for irrigation.
- **Irrigation Methods**: The choice of irrigation method (e.g., surface irrigation, drip irrigation, sprinkler systems, or rain-fed systems) is essential for efficient water use. Drip irrigation, for instance, conserves

water and minimizes disease risks by delivering water directly to plant roots.

- **Water Rights and Legal Restrictions**: In many areas, water usage is regulated by the government, and it's important to verify whether the project can legally access the required water resources.

Example:

For greenhouse vegetable production, drip irrigation is often the best choice because it uses less water, reduces weed growth, and minimizes the risk of fungal diseases. It's essential to ensure that water pressure is sufficient, and the filtration system is in place to prevent clogging of the drip lines.

Conclusion

Technical feasibility is a vital aspect of ensuring that an agricultural project is not only viable but also practical and sustainable. By assessing the suitability of land and soil, ensuring reliable water access, and verifying the availability of necessary resources, farmers and project planners can ensure that the conditions for success are in place before they proceed. The next sections will dive into the other key components, such as evaluating equipment, technology, and human resources, that further contribute to the technical feasibility of an agricultural project.

6.4.3 Climate and Weather Suitability

Climate is a critical factor in determining whether an agricultural project can thrive in a specific location. Every crop or livestock breed has specific temperature, rainfall, humidity, and other climatic requirements. A detailed assessment of local climate conditions ensures that the project will match the biological needs of the enterprise.

Important Parameters to Assess:

- **Temperature Ranges (min/max thresholds):** Most crops and livestock have optimal temperature ranges for growth and reproduction. If the temperature exceeds these thresholds (either too high or too low), it can affect yield, quality, or even survival. For example, tropical crops like bananas require warm temperatures, while temperate crops like wheat require cooler conditions.

- **Rainfall Quantity and Distribution:** Adequate and well-distributed rainfall is crucial for crop growth, especially for rainfed agriculture. Excessive rainfall can lead to waterlogging, while insufficient rainfall can cause drought stress. The seasonality of rainfall, including periods of dry spells, must also be assessed to determine whether irrigation systems will be needed.
- **Humidity and Wind Patterns:** Excess humidity can increase the risk of fungal diseases, while low humidity can stress crops. Wind speed and patterns are important to assess, as strong winds can cause damage to crops or create evaporation stress. Windbreaks may be needed in areas with high wind exposure.
- **Frost or Drought Risk:** Frost can be damaging to crops, particularly fruit trees and vegetables, while prolonged droughts can threaten both crops and livestock. An understanding of the risk of these extremes helps in planning for mitigation, such as frost protection methods or drought-resistant varieties.
- **Season Length and Growing Degree Days:** The length of the growing season (the period between the last frost and the first frost) and the cumulative temperature (measured in growing degree days) help determine the appropriate crops to grow and whether the project can meet its yield targets within the available growing period.

Tools for Climate Assessment:

- **Meteorological Data (Past 5–10 years):** Historical weather data provides insights into temperature, rainfall, and other climatic variables that influence the project.

- **Agro-Climatic Zoning:** This involves dividing the land into zones based on climatic conditions to assess which crops or livestock will thrive in each zone.
- **Crop-Climate Suitability Matrices:** These matrices match crops with the most appropriate climate conditions, allowing project planners to choose varieties that are suited to the local climate.

Example: Maize typically requires 500–800 mm of rainfall evenly distributed over a 120-day growing season. In regions with unpredictable rainfall, maize cultivation might be impractical unless supplemented with irrigation or dryland farming techniques.

6.4.4 Selection of Crop Varieties or Livestock Breeds

The selection of the right crop varieties or livestock breeds is fundamental for the success of an agricultural project. Choosing varieties that are well-adapted to local conditions, meet market demands, and offer high productivity is crucial.

Criteria for Selection:

- **Yield Potential:** The variety should be capable of high yields under local conditions. It is essential to consider not only the genetic potential of the variety but also how it will perform in the local environment, including soil and climate.
- **Disease and Pest Resistance:** Varieties or breeds resistant to common pests and diseases in the region will require fewer chemical interventions, reducing input costs and environmental impact.
- **Maturity Period:** The growing season and maturation rate of a variety should align with the local climate and market demand. Fast-maturing varieties may be more suitable for short growing seasons or high-value markets, while longer-maturing varieties may be more appropriate for crops with stable market demand over extended periods.

- **Quality Traits:** Some crops or breeds may be selected based on specific quality traits that meet consumer preferences or market demands. For example, high sugar content in sugarcane, higher oil yield in oilseeds, or high protein content in livestock breeds.
- **Local Acceptance and Market Preferences:** It is important to consider whether the local community accepts the chosen varieties or breeds, and whether they meet the needs of the target market. This may involve consumer preferences for taste, color, size, or specific certifications like organic or fair trade.

Example: In dryland areas, choosing drought-tolerant varieties like sorghum or early-maturing maize may be more viable than traditional maize varieties, as they can survive and yield better under low-water conditions.

6.4.5 Equipment and Infrastructure Requirements

Agricultural projects often require specialized equipment and infrastructure to ensure efficiency and productivity. This includes farm machinery, irrigation systems, storage facilities, processing units, and transportation networks.

Common Equipment and Infrastructure Requirements:

- **Farm Machinery:** Modern farming techniques often rely on mechanized equipment like tractors, planters, and harvesters to increase efficiency, reduce labor costs, and improve yields. The size and type of machinery should match the scale of the farm and the available land.
- **Irrigation Systems:** For projects relying on irrigation, the selection of appropriate systems (e.g., drip, sprinkler, or surface irrigation) is crucial. These systems need to be compatible with local water availability and the crops' needs.
- **Storage Facilities:** Adequate storage infrastructure such as silos, cold rooms, or warehouses ensures that harvested products are protected from spoilage and pests, especially for perishable goods.

- **Processing Units:** Some agricultural projects involve processing activities (e.g., milling, grading, or packaging) that require dedicated facilities. This adds value to raw products, extends shelf life, and opens up markets for processed goods.
- **Transport and Access Roads:** Access to reliable transport systems and roads is necessary to move raw materials, inputs, and finished products to and from the market.

Analysis Considerations:

- **Availability of Equipment Locally:** Assess whether the required equipment is readily available in the local market or whether it will need to be imported.
- **Cost of Purchase vs. Leasing:** Depending on the scale of the project, purchasing equipment may be costly. Leasing options may be viable, particularly for expensive machinery like harvesters or processing units.
- **Maintenance and Spare Parts Supply:** The availability of spare parts and maintenance services is important to avoid long downtime in case of equipment breakdowns.
- **Operator Training and Safety:** Skilled operators are needed to ensure that equipment is used efficiently and safely. Training programs for operators may be required to reduce risks and improve operational efficiency.

6.4.6 Input Supply and Management

For an agricultural project to be technically feasible, there must be a reliable and timely supply of essential inputs. Managing input supply ensures that crops and livestock are well cared for throughout their life cycle.

Key Inputs to Evaluate:

- **Certified Seeds or Planting Material:** The quality and certification of seeds or planting material determine the success of the crop. Using certified seeds ensures high germination rates, disease-free stock, and superior genetic potential.

- **Fertilizers and Soil Amendments:** Proper soil nutrition is vital for healthy crop growth. Fertilizers and soil amendments need to be assessed for their availability, quality, and cost.
- **Agrochemicals (Herbicides, Fungicides, Insecticides):** To protect crops from pests and diseases, reliable access to safe and effective agrochemicals is necessary.
- **Feed and Veterinary Supplies (for Livestock):** Livestock projects require constant access to high-quality feed, supplements, and veterinary supplies for animal health.
- **Packaging Materials:** Packaging is critical for transporting produce and ensuring it reaches the market in good condition. Quality packaging helps in preserving produce and adds value to the product.

Risk Considerations:

- **Price Volatility:** Input prices can fluctuate due to seasonal demand or supply chain disruptions. Price volatility can affect the project's profitability.
- **Fake or Low-Quality Inputs:** The use of substandard inputs (e.g., seeds, fertilizers) can lower productivity and lead to crop failures.
- **Delivery Delays or Shortages:** Delays in the delivery of critical inputs, especially during peak seasons, can cause delays in planting or harvesting, affecting yield.

6.4.7 Human Resource and Technical Skills

The availability of skilled labor is essential for implementing an agricultural project successfully. The workforce must be capable of handling the various technical aspects of the project, from planting to harvesting, irrigation, and equipment maintenance.

Required Skills:

- **Agronomic Practices:** Knowledge of planting techniques, crop nutrition, pest control, and harvesting methods is necessary to optimize production.
- **Livestock Management and Animal Health:** For livestock projects, expertise in animal care, feeding, breeding, and health management is crucial to maintaining productivity.
- **Equipment Operation and Repair:** Operators must be skilled in using and maintaining farm machinery to ensure it operates effectively.
- **Recordkeeping and Data Collection:** Accurate records of production, costs, and other key metrics are essential for managing the project and making informed decisions.
- **Quality Control and Post-Harvest Handling:** Ensuring that produce meets market quality standards and handling it correctly after harvest is crucial for reducing waste and maximizing profit.

Solutions:

- **On-the-Job Training:** This can be provided to workers to develop the necessary skills for their specific roles.
- **Hiring External Consultants:** For specialized tasks or knowledge gaps, external consultants can offer valuable expertise, particularly during the initial stages.
- **Partnerships with Extension Services or Universities:** Collaboration with agricultural extension services or academic institutions can provide access to up-to-date research, training, and advice.

Conclusion

The technical feasibility analysis evaluates the resources, infrastructure, and expertise necessary for an agricultural project to succeed. By carefully considering climate, soil, water, inputs, technology, and labor, project planners can identify potential barriers to success and devise solutions to address them.

This phase ensures that the project is not only feasible on paper but also executable under real-world conditions. The next chapter will explore financial feasibility, assessing whether the project can generate sufficient returns to be sustainable.

6.5 Developing a Technical Implementation Plan (TIP)

A **Technical Implementation Plan (TIP)** is a comprehensive roadmap that outlines how all the technical components of an agricultural project will be coordinated and executed. It ensures that every aspect of the project, from land preparation to harvesting, operates smoothly and efficiently. A TIP should be developed early in the planning process to ensure that technical feasibility aligns with overall project goals.

Key Elements of a TIP:

1. **Detailed Activity Calendar (Aligned with Seasons):** A well-structured activity calendar outlines the timeline for each key phase of the project. It should consider seasonal variations such as planting and harvesting windows, irrigation schedules, fertilization, pest control, and labor availability. The calendar helps synchronize activities with the optimal conditions for growth (e.g., planting crops before the rainy season or aligning harvests with market demand).

   Example:

   - **Month 1:** Soil testing and land preparation
   - **Month 2–3:** Planting season (depending on the crop)
   - **Month 4–6:** Irrigation and fertilizer application
   - **Month 7–9:** Harvest and post-harvest handling

2. **Input Procurement Schedule:** This component outlines when and how key agricultural inputs (seeds, fertilizers, pesticides, etc.) will be acquired. It ensures that all necessary inputs are available when needed to avoid

delays. The schedule also factors in the lead time required for ordering and delivery, considering local supply chain dynamics.

Example:

- **Month 1:** Order certified seeds, fertilizers, and irrigation systems
- **Month 2:** Ensure the delivery of seeds and machinery before planting season begins

3. **Staffing and Training Plan:** A comprehensive staffing plan identifies the number and type of personnel required at different stages of the project, from field workers to supervisors. It also includes a training plan to equip the staff with the necessary skills, such as operating machinery, applying fertilizers safely, or implementing pest management strategies. Training should align with the project's technical requirements to ensure smooth operations.

   Example:

   - **Month 1:** Hire farm managers and seasonal labor
   - **Month 2:** Train workers on crop management, irrigation, and pest control
   - **Month 3:** Continuous on-the-job training during the growing season

4. **Equipment Deployment and Usage Schedule:** The TIP should specify when and how each piece of equipment will be used throughout the project. It also details the maintenance schedule to ensure that machinery and tools are in optimal condition. Equipment deployment must be synchronized with field activities, such as plowing, planting, irrigation, and harvesting.

   Example:

   - **Month 1:** Deploy tractor for land preparation
   - **Month 2:** Use planters for sowing
   - **Month 3:** Implement drip irrigation system

5. **Monitoring Framework (e.g., Production Benchmarks):** A monitoring framework helps track progress toward production targets and identifies

any deviations from the plan. Benchmarks are used to evaluate crop growth, pest incidence, water usage, and overall project performance. Regular monitoring ensures that adjustments can be made before issues become major problems.

Example:

- **Monthly:** Monitor crop growth and health (e.g., leaf color, size, or disease signs)
- **Quarterly:** Assess irrigation efficiency and soil moisture levels

6.6 Technical Sustainability Considerations

Technical sustainability ensures that agricultural projects are not only productive but also environmentally responsible, socially beneficial, and economically viable over the long term. Implementing sustainable practices enhances productivity while minimizing negative impacts on the environment.

Key Sustainability Considerations:

1. **Use Conservation Agriculture Techniques:** Conservation agriculture (CA) focuses on minimizing soil disturbance and optimizing crop production through sustainable methods. It involves practices like minimal tillage, which preserves soil structure and reduces erosion. It also includes the use of cover crops, which prevent soil erosion, improve soil fertility, and promote biodiversity.

   Example:

   - **Minimum Tillage:** Reduces soil erosion and improves moisture retention
   - **Cover Crops:** Adds organic matter to soil, enhancing fertility and protecting it from wind and water erosion

2. **Apply Integrated Pest Management (IPM):** Integrated Pest Management is an environmentally sensitive approach to pest control that

combines biological, cultural, mechanical, and chemical practices. IPM aims to minimize pesticide use, reduce environmental contamination, and promote biodiversity. It helps ensure that pest control methods are effective, sustainable, and cost-efficient.

Example:

- **Biological Control:** Introducing natural predators of pests (e.g., ladybugs to control aphids)
- **Cultural Practices:** Rotating crops or selecting pest-resistant varieties to reduce pest buildup

3. **Choose Renewable Energy Options Where Possible:** Renewable energy sources, such as solar, wind, or biogas, can significantly reduce an agricultural project's dependence on non-renewable energy and lower operational costs. For instance, solar-powered irrigation systems can replace diesel-powered pumps, reducing fuel costs and promoting a cleaner environment.

Example:

- **Solar Irrigation Systems:** Reduce electricity costs and environmental impact by using solar energy to power water pumps
- **Wind Energy:** Use wind turbines for electricity generation, especially in areas with high wind potential

4. **Design for Climate Resilience:** Climate resilience ensures that agricultural projects can withstand the impacts of climate change, such as droughts, floods, or temperature extremes. Designing for resilience involves selecting climate-adapted crops or breeds, using water-efficient irrigation systems, and building infrastructure that can withstand extreme weather events.

Example:

- **Drought-Tolerant Crops:** Grow varieties like sorghum or millet in arid regions to mitigate water scarcity

- **Raised Beds in Flood-Prone Areas:** Prevent root rot and soil erosion by elevating crops in flood-prone zones

6.7 Challenges in Technical Feasibility

Despite the best plans, agricultural projects face various technical challenges that can hinder their success. Identifying these challenges early allows for proactive solutions and mitigates the risk of failure.

Common Technical Feasibility Challenges:

Challenge	Example
Inappropriate Technology	Using machinery too large for smallholder plots
Lack of Skilled Labor	Farmers not trained in drip irrigation or spraying safety
Infrastructure Gaps	No storage or cold chain to handle perishable produce
Environmental Constraints	Land too steep for mechanization or prone to erosion

1. **Inappropriate Technology:** Sometimes, the technology chosen for the project may not be suitable for the specific conditions of the farm, such as machinery that is too large for small plots or high-tech systems that require too much capital.

   Example:

   - Using large tractors on smallholder farms where manual labor or smaller machines would be more efficient and cost-effective.

2. **Lack of Skilled Labor:** Without adequately trained workers, even the best equipment and practices can fail. Lack of knowledge in specialized techniques, such as drip irrigation or pesticide safety, can lead to poor results and wasted resources.

Example:

- Farmers not trained in safe pesticide application, leading to poor crop protection and potential health risks.

3. **Infrastructure Gaps:** Inadequate infrastructure, such as poor road networks, lack of storage facilities, or absent cold chains, can severely affect the efficiency of an agricultural project. For example, perishable crops may spoil before reaching the market if refrigeration or transportation infrastructure is not available.

   Example:

   - No cold storage facility for fruits, leading to post-harvest losses due to spoilage.

4. **Environmental Constraints:** Physical challenges, such as unsuitable land for mechanization or areas prone to erosion, can limit the productivity of agricultural projects. This may require adjusting the approach or implementing additional measures to mitigate these constraints.

   Example:

   - Land that is too steep for mechanized farming equipment or that suffers from frequent soil erosion, requiring additional conservation practices.

6.8 Summary

Technical feasibility is essential to ensuring that an agricultural project is practically executable, considering factors like land, climate, inputs, equipment, and labor. A detailed technical analysis helps connect market demand with what is achievable in terms of productive capacity. By identifying constraints and technical challenges early, project planners can devise solutions, such as selecting the right technology, training workers, and investing in infrastructure. The next chapter will shift focus to **Financial Feasibility**, where we assess the profitability and long-term financial viability of the project.

Chapter 7: Financial Feasibility

Benefit:

Develops core competencies in budgeting, cost analysis, and financial planning. Essential for business plan development and investment analysis. Researchers can apply financial modeling to compare farming systems or innovations.

7.1 Introduction

A technically sound and market-ready agricultural project still needs to answer a crucial question: Will it make financial sense? Financial feasibility assesses whether the project will generate enough income to cover costs, repay loans, provide returns to investors, and sustain itself over time. This chapter explains how to estimate costs, forecast revenues, evaluate profitability, and determine funding strategies for agricultural projects.

7.2 What is Financial Feasibility?

Financial feasibility refers to the process of determining whether an agricultural project can generate enough revenue to cover its costs and provide an adequate return on investment (ROI) over its lifespan. Essentially, it involves assessing whether the project will be profitable, financially viable, and sustainable in the long term.

In agricultural projects, financial feasibility is particularly important because farming is subject to a variety of unpredictable factors, such as seasonality, price volatility, and climate risks. These factors can greatly impact both operational costs and revenues, so financial feasibility analysis must take these elements into account.

Key considerations for financial feasibility include:

- **Capital and operational costs**: How much it will cost to set up and run the project.

- **Revenue projections**: What income the project is expected to generate based on market demand, product pricing, and yield forecasts.
- **Profitability analysis**: Determining whether the project will generate profits after covering all costs.
- **Risk assessment**: Accounting for potential risks like crop failures, price fluctuations, and climatic challenges.
- **Funding**: Evaluating the financial requirements of the project and how these will be met (through loans, equity, grants, etc.).

7.3 Objectives of Financial Feasibility Analysis

The main objectives of financial feasibility analysis are to ensure that the agricultural project is not only operationally viable but also financially sound. Key goals include:

1. **Estimate Total Capital and Operational Costs:**
    - Accurately estimating the total costs involved in starting and running the project is crucial. These include both one-time capital investments and ongoing operational expenses.
2. **Forecast Revenues Based on Market Analysis:**
    - Based on market demand and production capacity, forecasting revenues helps assess the potential income from the sale of the products. This forecast must be adjusted for price fluctuations, seasonal changes, and any risks affecting the market.
3. **Conduct Profitability Analysis:**
    - Analyzing profitability involves comparing projected revenues with total costs to ensure that the project will generate sufficient returns. Profitability indicators like the gross margin and net profit margin are key measures.

4. **Evaluate Investment Return Metrics:**
    - Calculating key financial metrics such as Return on Investment (ROI), Net Present Value (NPV), and Internal Rate of Return (IRR) helps assess whether the project is financially worthwhile over time.
5. **Assess Cash Flow Needs and Funding Gaps:**
    - A detailed cash flow analysis ensures that the project will have enough liquidity to cover expenses, particularly during off-season periods when revenues may be low. Identifying any funding gaps helps in securing the necessary financial resources.
6. **Explore Financing Sources and Repayment Capacity:**
    - Financial feasibility also involves identifying potential funding sources (e.g., loans, grants, investors) and assessing the project's ability to repay these funds within a reasonable timeframe. The ability to repay loans, for example, is crucial for long-term financial sustainability.

7.4 Cost Estimation

Cost estimation is a critical part of financial feasibility analysis as it helps determine how much capital is needed and the expected operational costs. Costs in agricultural projects are generally categorized into two main groups: capital costs (fixed costs) and operational costs (recurring or variable costs).

7.4.1 Capital Costs (Fixed Costs)

Capital costs are one-time expenses required to set up the agricultural project. These costs are generally incurred before production begins and typically involve large investments. Once paid, these costs do not recur during the life of the project (unless there is a need for replacement or expansion).

Examples of Capital Costs:

1. **Land Acquisition or Leasing:**
   - Purchasing or leasing land for the agricultural project is often the largest capital cost. The location, size, and quality of land directly affect the viability of the project.
2. **Construction (Greenhouses, Storage, Processing Units):**
   - If the project involves greenhouses or storage facilities (such as silos or cold rooms), construction costs need to be included. Processing units may also be needed if value-added products are part of the business plan.
3. **Equipment and Machinery:**
   - This includes tractors, plows, planters, harvesters, and other mechanization equipment. It may also include specialized machinery like irrigation systems, sorting or packing lines, and processing machinery.
4. **Vehicles:**
   - Transport vehicles, such as trucks or farm-specific vehicles, are necessary for moving goods, equipment, or produce. The costs of purchasing or leasing vehicles should be estimated.
5. **Initial Infrastructure (Irrigation Systems, Fencing):**
   - Any initial infrastructure, such as setting up irrigation systems, installing fences, or building access roads, should also be considered.

Example:

A large-scale avocado farm might have capital costs of $500,000, including land acquisition, greenhouse construction, irrigation system installation, and purchasing tractors and harvesters.

7.4.2 Operational Costs (Recurring or Variable Costs)

Operational costs are the recurring expenses required to run the agricultural project throughout its life cycle. These costs fluctuate based on the scale of production, market prices, and seasonality. They are essential for day-to-day operations and must be carefully estimated to ensure the project remains financially sustainable.

Examples of Operational Costs:

1. **Seeds, Fertilizers, Agrochemicals:**
   - Costs for seeds, fertilizers, pesticides, herbicides, fungicides, and other inputs required to grow crops or manage livestock.
2. **Labor (Hired Workers, Technical Staff):**
   - Wages for farm workers, technical staff (e.g., agronomists), farm managers, and seasonal labor during planting, harvesting, and other critical periods. Labor costs can vary depending on the scale of the operation and the complexity of tasks.
3. **Fuel and Energy:**
   - Fuel costs for tractors, generators, irrigation pumps, and other machinery. Energy costs may include electricity or solar power for operations.
4. **Water Usage:**
   - Costs associated with accessing and using water for irrigation, particularly in areas where water is not abundant or must be purchased.
5. **Transport and Marketing:**
   - Costs to transport products from the farm to the market or distribution centers, including the cost of packaging, labeling, and transportation logistics.

6. **Maintenance and Repairs:**
    - Routine maintenance and repairs for machinery, irrigation systems, vehicles, and other equipment used in the operation.

Example:

For a tomato farm, operational costs might include:

- $20,000 annually for seeds and fertilizers,
- $30,000 for labor costs,
- $10,000 for fuel and irrigation maintenance,
- $15,000 for marketing and transportation.

It's important to distinguish between startup costs (which are incurred during the initial phases, such as land preparation and equipment purchases) and recurring costs (which are incurred on a regular basis throughout the farming cycle, such as labor, seed, and fertilizer costs).

Summary of Financial Feasibility Analysis

Financial feasibility is a vital aspect of any agricultural project, determining if it can generate sufficient returns to cover costs and deliver a profit. By estimating both capital costs (for setup) and operational costs (for ongoing production), project planners can create a clear financial roadmap. Forecasting revenues, evaluating risks, and determining funding requirements ensure that the project can be executed without financial strain. Proper financial analysis also prepares the project to adapt to market volatility, price fluctuations, and external risks, helping to maintain financial sustainability over time.

7.5 Revenue Projections

Revenue projections are an essential part of financial feasibility analysis, as they allow the project to estimate how much money it can generate over time. Revenue is calculated as the quantity of products sold multiplied by the price per unit. This projection is influenced by various factors, including expected yields, market prices, product grades, and sales channels.

Key Factors Influencing Revenue:

- **Expected Yields**: Yield predictions are derived from technical feasibility assessments (e.g., soil quality, climate, and crop type) and reflect how much of a given product (such as crops or livestock) can be produced in a specific time frame. Yields directly impact the volume of product available for sale and, consequently, the revenue generated.
- **Market Prices**: The prices that can be obtained for the product in the market are a key determinant of revenue. Prices can fluctuate based on demand, market conditions, competition, and other factors outlined in the market feasibility study. Understanding price trends and market dynamics helps in projecting realistic revenue.
- **Product Grades**: In many agricultural sectors, products are graded based on quality. For example, premium-quality tomatoes or organic produce may fetch a higher price than standard-grade products. These grade distinctions should be considered when projecting revenue to reflect potential price variations.
- **Sales Channels**: The way products are sold (e.g., wholesale, retail, export) affects pricing and volumes. Wholesale prices tend to be lower but may result in higher volume sales, while retail or export markets may offer higher prices but lower volumes.

Example:

Let's say a farm is planning to produce 50 tons of tomatoes annually and expects to sell at $400 per ton.

- **Expected Revenue** = 50 tons × $400 = **$20,000 per year**.

This straightforward calculation helps provide a baseline for expected income. However, for more complex projects, other factors like seasonal price fluctuations, product diversification (e.g., selling both fresh and processed tomatoes), and market positioning can also influence projections.

7.6 Profitability Analysis

Profitability analysis helps evaluate whether the project will generate enough profit to justify the investment. Several financial metrics and tools are used to assess profitability and financial performance:

7.6.1 Break-Even Analysis

Break-even analysis helps determine the minimum output (or price) required to cover all costs, which means there is no profit or loss at this point. This analysis is crucial for understanding at what level of production or sales the project will start to generate profits.

The formula to calculate break-even volume is:

- **Break-even Volume** = Fixed Costs / (Selling Price – Variable Cost per Unit)

Explanation:

- **Fixed Costs** are those that do not change with the level of production, such as rent, machinery, and salaried employees.
- **Selling Price** is the price at which the product is sold.
- **Variable Cost per Unit** refers to costs that vary directly with the level of production, such as raw materials, labor costs, and water or fuel.

For example, if the project has fixed costs of $10,000 and the selling price is $400 per ton, with a variable cost of $250 per ton:

- **Break-even Volume** = $10,000 / ($400 – $250) = **40 tons**

This means the project must produce and sell at least 40 tons to cover all fixed and variable costs. Any production above this will result in profit.

7.6.2 Net Present Value (NPV)

Net Present Value (NPV) is a financial metric used to evaluate the profitability of a project by calculating the present value of expected future cash flows, discounted by a rate that reflects the project's risk or cost of capital.

How NPV Works:

- NPV helps determine whether a project is worth investing in by considering the time value of money. Future revenues and costs are discounted back to present value terms, allowing investors to compare cash flows that occur at different times.
- If **NPV > 0**, the project is considered financially viable, as it is expected to generate more cash than it costs to fund.
- The **higher the NPV**, the more attractive the investment becomes because it indicates higher profitability over time.

Example:

If the project is expected to generate cash flows of $50,000 in year one, $60,000 in year two, and $70,000 in year three, and the discount rate is 10%, the NPV calculation would involve discounting those cash flows to present value and summing them up. If the total discounted cash flow exceeds the initial investment, the NPV is positive.

7.6.3 Internal Rate of Return (IRR)

The **Internal Rate of Return (IRR)** is the discount rate that makes the NPV of the project equal to zero. It is used to assess the expected rate of return on the project.

- **If IRR > cost of capital**, the project is considered financially viable because the return exceeds the cost of financing the project.
- **The higher the IRR**, the more attractive the investment, as it indicates the project generates a higher rate of return on investment compared to the cost of capital.

For example, if the project is expected to generate annual cash flows of $30,000 for five years, and the IRR is calculated at 15%, the project is deemed viable if the cost of capital is 10%.

7.6.4 Payback Period

The **payback period** refers to the amount of time required to recover the initial investment from the project's profits.

- **Shorter Payback Period** means lower investment risk, as the project will recover its costs more quickly.
- **Longer Payback Period** might indicate higher risk, especially if there are uncertainties related to future revenue generation.

A project with a payback period of two years means the initial investment will be repaid within two years from profits. This is an important metric for investors who want to know how long it will take for them to get their money back.

7.6.5 Benefit-Cost Ratio (BCR)

The Benefit-Cost Ratio (BCR) is used to compare the total expected benefits to the costs of the project, helping assess whether the project will deliver value. The formula is:

- **BCR** = Present Value of Benefits / Present Value of Costs

A **BCR > 1.0** indicates a profitable investment, meaning that the benefits outweigh the costs. A BCR less than 1.0 suggests the project is not financially viable, as the costs exceed the expected benefits.

7.7 Cash Flow Forecasting

Cash flow forecasting is essential for ensuring that the project remains liquid and can meet its financial obligations over time. Cash flow refers to the timing of inflows (revenue) and outflows (expenses) and is crucial for maintaining day-to-day operations.

Key Elements of Cash Flow Forecasting:

- **Monthly or Quarterly Inflows and Outflows**: Projecting cash flow on a monthly or quarterly basis ensures that the project will have enough liquidity to cover expenses, such as payroll, loan repayments, or supplier payments.
- **Seasonal Peaks in Expenses**: Agriculture is subject to seasonal fluctuations, and expenses may peak during planting or harvesting seasons. Forecasting these peaks helps ensure that adequate funding is available at those critical times.
- **Time Lags Between Production and Income**: For many agricultural products, there is a time lag between production and sales (e.g., growing seasons, harvest-to-sale periods). Proper cash flow forecasting must account for this lag to prevent liquidity shortages.
- **Loan Repayments and Interest Schedules**: If external financing is used, the cash flow forecast should include loan repayment schedules to ensure the project can meet its debt obligations without financial strain.

Example:

If the farm expects $100,000 in revenue in months 4 through 6, but needs $70,000 for planting costs in months 1 and 2, the project should have sufficient liquidity to cover these early expenses. Proper cash flow management ensures that the farm will have enough funds to maintain operations through the entire cycle.

Summary

Revenue projections, profitability analysis, and cash flow forecasting are essential components of the financial feasibility of an agricultural project. By using tools like break-even analysis, NPV, IRR, and BCR, project planners can evaluate whether the project is financially viable. Ensuring the project has sufficient liquidity through accurate cash flow planning is critical for avoiding financial pitfalls, especially in seasonal industries like agriculture.

7.8 Sensitivity and Risk Analysis

Sensitivity and risk analysis are critical components of a feasibility study, as they test a project's ability to remain viable under varying conditions. These analyses help identify which variables have the most significant impact on project outcomes and assist in risk mitigation planning.

Sensitivity Analysis

This involves systematically altering key assumptions—such as prices, yields, and costs—to assess their effect on project profitability and sustainability.

Variable	Impact if Changed
Selling Price ↓ 10%	A reduction in selling price can significantly erode profit margins, especially in high-input systems. In some cases, it may push the net return into negative territory.
Yield ↓ 20%	A substantial drop in yield may lead to operational losses, especially if fixed costs remain constant. This is particularly risky in monoculture systems or regions with high climate variability.
Input Prices ↑ 15%	Higher input prices (seeds, fertilizer, fuel, labor) raise the break-even point. If not matched by increased output or prices, profitability may be compromised.
Loan Interest ↑ 3%	An increase in loan interest raises the debt servicing burden and can affect cash flow and net income, particularly in capital-intensive projects.

Scenario Analysis

This complements sensitivity analysis by modeling different combinations of variable changes under three primary scenarios:

- **Best-Case Scenario**: Assumes optimal conditions (e.g., high yields, stable prices, low input costs). Helps assess the maximum potential return.
- **Base-Case Scenario**: Reflects the most likely set of assumptions. Used as a benchmark.
- **Worst-Case Scenario**: Assumes adverse conditions (e.g., price drop, poor yield, rising costs). Helps determine if the project can withstand financial stress.

Conducting these analyses provides insight into the project's risk exposure and guides strategic planning, such as setting up reserves, diversifying production, or purchasing insurance.

7.9 Sources of Financing

Financing is the lifeblood of any agricultural project. The choice of financing method affects both the capital structure and long-term sustainability of the venture. Below are the main categories of financing sources:

Equity Capital

Equity funding does not require repayment and often comes from internal or partner resources:

- **Personal Savings**: Often used in small to medium-scale projects. Demonstrates owner commitment.
- **Partner/Investor Contributions**: Attracting investors can provide significant capital infusion but may dilute control.

Debt Financing

Loans provide leverage but require disciplined repayment, typically with interest:

- **Commercial Banks**: Offer a variety of term loans and credit lines, but may require collateral.
- **Microfinance Institutions**: Suitable for smallholder or cooperative-based projects, often with more flexible terms.
- **Government Agricultural Loans**: Frequently subsidized with favorable interest rates and grace periods. Require adherence to government regulations.

Grants and Subsidies

Non-repayable funds from government or non-governmental organizations:

- **Government Rural Development Programs**: Often target infrastructure, mechanization, or specific value chains.
- **International Donors/NGOs**: Provide funding for sustainable, climate-resilient, or socially inclusive agricultural projects.

Blended Finance

A strategic mix of grants, concessional loans, and private investment designed to reduce investment risk and attract commercial financing:

- Used in large-scale or innovative agri-projects.
- Enhances financial sustainability and development impact.

Note: Always ensure that **debt servicing obligations** are incorporated into **cash flow analysis**, and that repayment schedules align with the seasonal revenue cycle typical of agricultural enterprises.

7.10 Tools and Templates

Robust financial planning and monitoring require tools that can analyze, project, and track the financial health of agricultural projects. Project managers should utilize the following resources:

Excel-Based Tools

- **NPV (Net Present Value) and IRR (Internal Rate of Return) Calculators**: Help evaluate the financial viability of investments over time.
- **Break-Even Analysis Templates**: Determine the production volume or price required to cover all costs.
- **Budgeting Sheets**:
  - **CAPEX (Capital Expenditure)**: For long-term investments such as machinery, land, and infrastructure.
  - **OPEX (Operational Expenditure)**: For day-to-day running costs, including labor, maintenance, and inputs.

Loan Amortization Schedules

- Allow for tracking of loan repayment progress, including interest and principal components over time.

Cash Flow Trackers

- Help visualize and manage seasonal cash inflows and outflows, essential for agricultural projects with uneven revenue patterns.

Financial Dashboards

- Provide real-time visualization of key financial indicators (KPIs), such as gross margin, ROI, and cost-to-income ratio.
- Can be integrated with accounting software or mobile tools for ongoing project monitoring.

These tools must be **tailored to the project's size**, **commodity focus**, and **geographical context** to be effective. For instance, a greenhouse project in Kenya may require different budgeting assumptions than a rainfed rice project in Southeast Asia.

7.11 Summary

Financial feasibility is a critical aspect of evaluating agricultural projects, as it answers the question of whether the project will be profitable and financially sustainable over time. However, the process is not just about numbers and calculations; it involves careful consideration of the project's costs, revenues, risks, and ability to adapt to real-world conditions.

To evaluate **financial feasibility**, project planners rely on tools such as:

- **Cost estimation**: Accurate projections of both **capital costs** (e.g., equipment, land acquisition) and **operational costs** (e.g., inputs, labor) are necessary to determine the financial requirements of the project.
- **Revenue forecasting**: Based on market conditions, expected yields, and sales channels, revenue projections help assess whether the project can generate enough income to cover costs and provide a return on investment.
- **Profitability analysis**: This includes **break-even analysis** (which determines when the project will start to make a profit), **Net Present Value (NPV)**, **Internal Rate of Return (IRR)**, and **Payback Period** calculations. These metrics help determine whether the project will deliver a sufficient return on investment and how long it will take to recover initial costs.

However, financial planning also requires understanding:

- **Risk**: Agriculture projects are inherently risky due to factors like weather, disease, and fluctuating market prices. Financial feasibility analysis must account for these risks and create strategies to mitigate them.
- **Timing**: The timing of cash flows, from input costs to income from sales, is critical in determining whether the project can sustain operations over time.
- **Real-world conditions**: This includes potential issues like market instability, legal regulations, or environmental factors that could impact the project's success.

In summary, financial feasibility is essential for determining whether an agricultural project can stand the test of time and produce profits. It requires a deep understanding of the project's financial structure and the ability to adapt to external challenges. This chapter sets the stage for understanding economic and social feasibility, which looks at the broader impacts of the project on the community and society. The next chapter will delve into how agricultural projects can contribute to rural development, poverty alleviation, and food security.

Chapter 8: Economic and Social Feasibility

Benefit:

Promotes a holistic approach to project planning by integrating socioeconomic factors. Students can analyze community impact and inclusiveness, while researchers can conduct social return on investment (SROI) studies or gender-based evaluations.

8.1 Introduction

While financial feasibility focuses on the profitability and sustainability of a project for its investors, economic and social feasibility examines the broader impact of the project on the local community, economy, and society as a whole. This is especially important in the agricultural sector, where projects are often located in rural and underdeveloped areas, and the potential benefits extend far beyond the project itself.

Agricultural projects can be a powerful tool for economic development and poverty reduction. They can stimulate employment, improve the livelihoods of farmers and rural workers, and contribute to food security. However, these projects can also have negative externalities (unintended negative impacts on the environment, local communities, or public health), which need to be addressed in the planning process.

This chapter will explore how to evaluate the economic contributions and social implications of agricultural projects, focusing on their potential benefits, challenges, and ways to maximize positive outcomes while minimizing harm.

8.2 What is Economic and Social Feasibility?

Economic and social feasibility assesses how an agricultural project can contribute to:

- **Economic growth**: This includes evaluating whether the project will create jobs, stimulate local businesses, or contribute to regional economic development.
- **Social welfare**: This considers how the project might improve the quality of life for local communities, such as through better access to food, healthcare, and education, or through improved standards of living.
- **Equitable distribution of benefits**: It evaluates whether the benefits of the project will be fairly distributed, ensuring that vulnerable groups (e.g., smallholder farmers, women, youth) are not left behind.

In this context, **economic feasibility** focuses on the project's ability to generate wealth for the broader community, while **social feasibility** looks at how the project impacts society in terms of health, education, and overall well-being.

Economic Feasibility Considerations

1. **Job Creation**: Many agricultural projects create direct and indirect employment opportunities, such as farm labor, processing plant workers, transport, and distribution staff. The extent to which the project generates jobs, particularly for marginalized groups, is a key measure of economic feasibility.
2. **Supply Chain Development**: Agricultural projects often require the development of new infrastructure or the expansion of existing infrastructure (e.g., roads, storage facilities, and marketplaces). This can stimulate the local economy by fostering new business opportunities for local suppliers and service providers.
3. **Income Generation**: By providing a stable source of income to farmers and rural workers, agricultural projects can help alleviate poverty. The

project's potential to increase income levels and reduce poverty is a significant indicator of its economic impact.

4. **Infrastructure Development**: Agriculture often requires substantial infrastructure, such as irrigation systems, transport networks, and storage facilities. Assessing whether the project will improve or develop this infrastructure is an important part of the economic feasibility analysis.
5. **Food Security**: Projects that focus on increasing food production or improving access to affordable, nutritious food contribute to long-term food security. These projects reduce reliance on imports and help ensure that local populations have a stable supply of food.

Social Feasibility Considerations

1. **Livelihood Improvements**: How does the project improve the livelihoods of the local population? This includes assessing whether it creates better-paying jobs, enhances working conditions, or increases access to social services (healthcare, education).
2. **Social Equity**: Economic benefits should be distributed fairly across different segments of society, particularly marginalized groups like women, youth, and indigenous populations. Assessing how inclusive the project is and whether it promotes gender equality and youth empowerment is vital.
3. **Cultural Impacts**: Understanding the social fabric of the local community is crucial. Agricultural projects that change traditional practices (e.g., through the introduction of new crops or technologies) must consider cultural sensitivities and ensure that they align with local values and customs.
4. **Community Development**: Does the project contribute to broader community development, such as improving infrastructure, local governance, and education? This can include access to better healthcare,

schools, or community centers, which ultimately enhances the social stability of the region.

5. **Health and Well-being**: Agricultural projects can improve health outcomes by providing access to healthier food and promoting sustainable agricultural practices. However, they can also have negative health impacts, such as through pesticide use or poor working conditions, which need to be mitigated.
6. **Public Engagement and Participation**: In many cases, local communities should be involved in the planning and decision-making processes. This ensures that the project meets their needs and addresses any concerns they may have about its impacts.

Summary

In conclusion, economic and social feasibility assesses how an agricultural project can create benefits for the local community and contribute to the broader economy. It goes beyond just profitability for investors and evaluates how the project can improve livelihoods, stimulate local economies, and enhance social well-being. By considering issues such as job creation, equitable distribution of benefits, and cultural and health impacts, agricultural projects can become powerful tools for promoting sustainable development in rural areas. In the next chapter, we will explore Environmental Feasibility, which assesses the environmental impacts of the project and ensures that it is sustainable and ecologically responsible.

8.3 Objectives of Economic and Social Feasibility Analysis

Economic and social feasibility analysis aims to ensure that an agricultural project does not merely serve the interests of investors or developers, but also creates meaningful, positive, and sustainable impacts for society and the broader economy. This analysis seeks to:

- **Determine the project's impact on employment and income generation**:
  Agricultural projects, especially in rural areas, can provide much-needed employment opportunities—both permanent and seasonal. The analysis quantifies direct and indirect jobs created (e.g., farm laborers, transporters, traders) and evaluates how these translate into improved household income and regional economic activity.
- **Assess how the project contributes to poverty reduction and rural development**:
  Agriculture is a cornerstone of rural economies. Projects that enhance productivity, increase farmer income, or introduce infrastructure (like irrigation or roads) can directly contribute to lifting communities out of poverty. This objective examines the alignment of the project with poverty alleviation goals and the extent of its contribution to broader rural development indicators.
- **Evaluate whether the project is inclusive and socially acceptable**:
  A socially feasible project is one that considers the needs and voices of all stakeholder groups—especially marginalized populations such as women, smallholder farmers, indigenous communities, or landless laborers. This includes assessing participation in decision-making, access to benefits, and cultural sensitivity of interventions.
- **Analyze potential economic multipliers (value chain effects)**:
  Beyond immediate benefits, agricultural projects can create positive chain reactions throughout the economy. This includes job creation in input supply industries, expansion of agri-processing, and new business opportunities. The analysis quantifies and evaluates these multiplier effects to assess broader economic impact.

- **Identify social risks, including displacement or community conflict**: Projects may inadvertently lead to the displacement of communities, increased land disputes, or conflict over shared resources (e.g., water, grazing land). Feasibility analysis must identify and mitigate these risks to ensure social stability and minimize resistance from affected populations.
- **Align the project with local and national development plans**: Projects should not operate in isolation. This objective ensures that the project complements national and regional policies, such as agricultural development strategies, poverty reduction frameworks, or climate resilience plans, thereby enhancing policy coherence and increasing chances of public and donor support.

8.4 Economic Feasibility Components

Economic feasibility involves a comprehensive evaluation of a project's potential to contribute to economic growth and prosperity beyond its direct financial returns. It incorporates both tangible and intangible impacts, as detailed in the following components:

8.4.1 Cost-Benefit Analysis (CBA)

Cost-Benefit Analysis (CBA) in the context of economic feasibility differs from traditional financial analysis by incorporating social and economic values, including non-market benefits and externalities. It considers the project's overall contribution to societal welfare rather than just investor returns.

Key Elements:

- **Economic Benefits:**
  - **Job Creation (Direct and Indirect):** Employment across the value chain—from planting to processing to distribution—adds economic value and reduces unemployment.

- **Increased Food Production and Reduced Imports:** Boosting domestic output enhances food sovereignty, reduces foreign exchange burdens, and stabilizes food prices.
- **Value Addition through Processing:** Converting raw products (e.g., milk into cheese, tomatoes into paste) creates new revenue streams, supports agro-industries, and improves product shelf life.
- **Strengthened Local Markets and Trade:** By creating reliable supply and demand linkages, projects support market infrastructure, improve logistics, and boost rural trade.
- **Enhanced Tax Revenues (Indirect):** Formalized businesses contribute to the tax base, enabling governments to invest in further development initiatives.

- **Economic Costs:**
  - **Opportunity Cost of Land Use:** Land used for the project may displace other productive uses (e.g., grazing, forestry, or communal farming). These trade-offs must be weighed carefully.
  - **Public Investment:** Government support for infrastructure (roads, electricity, irrigation) and incentives (tax breaks, subsidies) represent costs to taxpayers and must be justified by long-term returns.
  - **Environmental Remediation Costs:** If the project causes ecological degradation (e.g., soil erosion, water pollution), these damages require restoration and rehabilitation—often at significant cost.

Example: A large-scale rice irrigation scheme may require major investments in canal infrastructure and displace small-scale farmers but could enhance national rice output, reduce imports, and create thousands of jobs in milling and transport.

8.4.2 Multiplier Effects

Agricultural projects often have **ripple effects** throughout the local and regional economy, known as **economic multipliers**. These effects extend far beyond the immediate beneficiaries of the project and can catalyze broader development.

Types of Multipliers:

- **Forward Linkages:**

  Projects supply raw materials to downstream industries such as:
  - Agro-processors (e.g., fruit to juice factories)
  - Exporters and packaging industries
  - Retailers and food distributors
- **Backward Linkages:**

  Demand for inputs, equipment, and services increases, stimulating local industries that supply:
  - Seeds, fertilizers, pesticides
  - Farm machinery and spare parts
  - Irrigation systems, construction firms, extension services
- **Spillover Impacts:**

  Projects can trigger:
  - **Local Infrastructure Development**: Improved roads, electricity, and water systems benefiting non-project residents.
  - **Business Growth**: New micro and small enterprises (e.g., shops, transport services, maintenance services) spring up to serve the growing economic activity.

Example: A dairy development project might lead to new fodder businesses, veterinary services, milk collection centers, and the development of rural refrigeration technologies.

8.4.3 Contribution to Food Security

Agricultural projects that enhance the availability, accessibility, and affordability of food contribute significantly to household and national food security.

Key Contributions:

- **Increased Local Food Supply:** By increasing agricultural productivity and extending growing seasons through irrigation or improved varieties, projects reduce food deficits and import reliance.
- **Stabilized Prices:** Reliable food supply helps smooth out seasonal price fluctuations and ensures affordability for low-income consumers.
- **Nutrition Improvement:** Projects that promote diversified farming systems (e.g., integrating vegetables, legumes, and small livestock) enhance dietary diversity and combat malnutrition.
- **Resilience to Shocks:** Strengthening food production systems increases resilience against external shocks such as droughts, economic crises, or conflicts that disrupt food imports.

Example: A project that promotes the cultivation of drought-tolerant millet and sorghum in arid areas not only improves local food availability but also buffers against climate-related yield failures.

8.5 Social Feasibility Components

Social feasibility examines the degree to which a project is acceptable and beneficial to the communities it affects. It considers cultural, demographic, and institutional factors that influence project success. A socially feasible agricultural project enhances equity, inclusion, and local empowerment while minimizing potential negative social consequences.

8.5.1 Community Acceptability and Cultural Fit

For a project to gain community support and long-term sustainability, it must align with the local context:

- **Cultural alignment**: Projects should respect local customs, traditions, and values. For example, introducing a livestock project in a community with religious restrictions on animal handling may provoke resistance.
- **Land tenure systems**: Understanding formal and informal land rights is crucial. Projects that overlook local land use norms can trigger disputes.
- **Gender norms**: Ignoring traditional gender roles in labor or leadership may cause social friction or reduce women's participation.

8.5.2 Stakeholder Engagement

Genuine participation of stakeholders throughout the project cycle fosters trust, ensures relevance, and builds local ownership:

- **Inclusive engagement**: Engage farmers, women's groups, youth associations, elders, and marginalized populations to understand diverse perspectives.
- **Communication**: Ensure information is accessible in local languages and formats (e.g., visual materials for low-literacy groups).
- **Grievance mechanisms**: Establish accessible, transparent systems for addressing complaints and concerns, ensuring timely resolution and community satisfaction.

8.5.3 Employment and Labor Conditions

Job creation is a key social benefit, but the quality and conditions of employment matter greatly:

- **Job types**: Projects should distinguish between permanent and seasonal employment, and promote decent work conditions across all roles.
- **Fair wages and safety**: Adherence to national labor standards on pay, working hours, and occupational health and safety is non-negotiable.

- **Skill development**: Capacity building—through training and apprenticeships—enables local people to advance into more skilled roles over time.

8.5.4 Gender and Youth Inclusion

Agricultural development is most impactful when it integrates traditionally underrepresented groups:

- **Women's empowerment**: Promote women's active roles not only as laborers but also as planners, leaders, and entrepreneurs.
- **Access equity**: Ensure women and youth have access to land, credit, training, and technologies.
- **Youth engagement**: Offer targeted programs in agripreneurship, ICT in agriculture, and modern farming techniques to make agriculture appealing to young people.

8.5.5 Impact on Vulnerable Groups

Projects must avoid unintended harm to those most at risk of exclusion or exploitation:

- **Displacement risks**: Large-scale land use changes may displace smallholders or indigenous communities.
- **Marginalization**: Technological upgrades or contract farming schemes might benefit wealthier farmers more than smallholders.
- **Environmental vulnerability**: Intensive input use may harm communities dependent on natural resources (e.g., through water pollution or soil degradation).

8.6 Tools for Economic and Social Analysis

A wide range of tools is available to assess both economic and social feasibility. Each offers insights into different dimensions of project performance:

Tool	Purpose
Cost-Benefit Analysis (CBA)	Measures overall economic value by comparing total costs and benefits.
Social Impact Assessment (SIA)	Evaluates how the project affects community well-being, equity, and cohesion.
Stakeholder Analysis	Identifies all key actors, their interests, and levels of influence.
Poverty Impact Assessment	Analyzes the extent to which the project can reduce poverty and inequality.
Gender Analysis Frameworks	Examines the different impacts of the project on men and women.
Community Consultations / Surveys	Gathers feedback on project design and social concerns from beneficiaries.

8.7 Environmental Sustainability Considerations

Economic and social feasibility are deeply intertwined with environmental sustainability. A project that depletes natural resources or contributes to environmental degradation is unlikely to be socially acceptable or economically viable in the long term.

Key environmental questions include:

- **Resource use**: Will the project deplete or sustainably manage soil fertility, water sources, and biodiversity?
- **Climate resilience**: Are climate-smart practices—such as conservation agriculture, drought-resistant crops, or agroforestry—being implemented?

- **Pollution control**: Are measures in place to mitigate greenhouse gas emissions, manage chemical use, and reduce waste?
- **Ecosystem impact**: Projects should avoid encroaching on protected areas or disrupting local ecosystems.

Failing to address environmental concerns can result in community opposition, regulatory sanctions, or loss of market access—particularly in export-oriented agriculture.

8.8 Aligning with Policy and Development Goals

Socially and economically feasible projects should also align with broader policy and development frameworks to attract funding, ensure policy support, and enhance scalability.

National and Regional Policy Alignment

- Projects should support national development strategies, including poverty reduction, food security, and rural transformation plans.
- Regional agricultural policies (e.g., Comprehensive Africa Agriculture Development Programme - CAADP) should also be considered where applicable.

UN Sustainable Development Goals (SDGs)

Projects aligned with SDGs demonstrate global relevance and long-term impact. Key relevant goals include:

- **SDG 1: No Poverty** – Through income generation and job creation.
- **SDG 2: Zero Hunger** – Via improved food availability and access.
- **SDG 5: Gender Equality** – By empowering women in agriculture.
- **SDG 8: Decent Work and Economic Growth** – Through dignified employment and agribusiness development.
- **SDG 13: Climate Action** – By integrating climate-smart agricultural practices.

8.9 Social Risks and Mitigation Strategies

Social risks are potential negative outcomes that may arise during the implementation of agricultural projects, especially if the social context is not properly understood or managed. These risks, if ignored, can jeopardize the success of the project, trigger opposition, or lead to long-term harm to communities. Identifying these risks early and developing appropriate mitigation strategies is a critical aspect of social feasibility.

Below is a breakdown of common social risks associated with agricultural projects, their potential impacts, and recommended mitigation measures:

Risk	Potential Impact	Mitigation Strategy
Land Disputes	Legal battles, project delays, community conflict	Conduct thorough land tenure assessments; secure documented consent from rightful landowners; involve traditional authorities and local governments.
Labor Exploitation	Reputational damage, legal penalties, workforce dissatisfaction	Establish and monitor fair labor standards; implement training on labor rights; conduct regular audits and promote worker representation.
Community Resistance	Project rejection, protests, operational disruption	Engage stakeholders from the outset; integrate community input into project design; maintain transparent communication and ongoing dialogue.
Gender Exclusion	Unequal distribution of	Apply gender-sensitive planning; offer dedicated support programs

	benefits, social inequity	for women; ensure representation in decision-making bodies.
Youth Marginalization	Loss of potential workforce and innovation; apathy	Design youth-targeted training, entrepreneurship initiatives, and technology-based interventions to make agriculture attractive.
Cultural Incompatibility	Misalignment with local beliefs or traditions	Conduct cultural assessments; adapt project components to respect local customs and values.
Displacement of Vulnerable Groups	Loss of homes or livelihoods for marginalized populations	Avoid land acquisition in sensitive areas; offer fair compensation, resettlement support, and livelihood restoration programs.
Environmental Degradation	Community health risks, resistance, ecological harm	Integrate environmental safeguards (e.g., buffer zones, pollution controls); ensure community awareness of mitigation plans.

Effectively addressing social risks requires not only technical solutions but also an empathetic, participatory approach that centers community wellbeing and fairness.

8.10 Summary

Economic and social feasibility assessments are indispensable for ensuring that agricultural projects deliver more than just financial returns. A truly successful agricultural initiative must foster:

- **Economic viability**: Generating profits, creating jobs, and boosting local economies.
- **Social inclusion**: Engaging diverse stakeholders, promoting equity, and respecting local values and traditions.
- **Long-term acceptability**: Building trust and ownership among communities to ensure continuity and sustainability.

When these dimensions are carefully analyzed and integrated into project design and management, agricultural ventures become powerful engines for inclusive rural development, poverty reduction, and social cohesion.

Such projects also build resilience—socially, economically, and environmentally - by fostering empowerment, reducing inequality, and preparing communities to adapt to changing conditions.

Key Takeaway:

A socially and economically feasible agricultural project is one that uplifts people, aligns with local values, and delivers enduring benefits across all levels of society.

What's Next?

In the next chapter, we turn our attention to **Environmental Sustainability Considerations**. This will involve examining how agricultural projects can balance productivity with the need to preserve natural ecosystems. Topics will include:

- Sustainable land and water use
- Climate-smart agricultural practices
- Biodiversity conservation
- Environmental risk mitigation

By bridging the social, economic, and environmental pillars, the following chapter will complete the triad of feasibility analysis that underpins responsible agricultural project management.

Part III: Sustainability, Monitoring, and Evaluation

Focus: Long-term project success, sustainability, and performance tracking.

Chapter 9: Environmental Sustainability Considerations

Benefits:

Chapter 9 equips students and researchers with essential knowledge on integrating sustainability into agricultural projects. It enhances understanding of environmental impacts, promotes responsible resource use, and supports the development of eco-friendly practices. The chapter also provides practical tools for assessing and improving environmental performance in project planning.

9.1 Introduction

Agriculture is a cornerstone of human survival, providing essential resources such as food, fiber, fuel, and raw materials. However, agricultural expansion and intensification have often come at a significant environmental cost. Common environmental consequences include deforestation, soil erosion, water pollution, greenhouse gas emissions, and loss of biodiversity. These impacts not only degrade ecosystems but also threaten the long-term viability of agricultural systems themselves.

In an era marked by climate change, resource scarcity, and growing global food demands, the sustainability of agricultural practices is more important than ever. Agricultural projects must, therefore, be designed with a strong environmental ethic—one that protects and regenerates the natural systems upon which agriculture depends.

This chapter explores the key environmental sustainability principles relevant to agricultural project planning and implementation. It outlines how to minimize ecological damage, adopt sustainable practices, and promote resilience and productivity for future generations.

9.2 What is Environmental Sustainability in Agriculture?

Environmental sustainability in agriculture refers to the implementation of farming practices that protect and enhance the environment while ensuring long-term agricultural productivity. It involves a harmonious interaction between agricultural activities and the natural ecosystem—preserving the integrity of soil, water, air, flora, and fauna.

Sustainable agriculture:

- Utilizes natural resources efficiently and responsibly,
- Minimizes pollution and environmental degradation,
- Enhances resilience to climate variability and change, and
- Promotes ecological balance and biodiversity.

In essence, it is a commitment to farming in a way that meets present food and fiber needs without compromising the ability of future generations to meet theirs.

9.3 Objectives of Environmental Sustainability Analysis

An environmental sustainability analysis in agricultural project planning serves to ensure that development objectives do not compromise environmental health. Key objectives include:

- Assessing the project's impact on natural resources, including soil quality, water availability, air quality, and local ecosystems.
- Ensuring farming systems are climate-resilient, able to withstand and adapt to shifting weather patterns and temperature extremes.
- Promoting biodiversity and conserving natural habitats, particularly those critical to pollinators, natural pest controllers, and soil microbes.
- Minimizing dependency on chemical inputs, favoring organic, biological, or integrated pest management strategies.
- Implementing sustainable water use practices, such as efficient irrigation and watershed management.

- Protecting and enhancing soil health, through erosion control, organic matter management, and responsible tillage practices.

9.4 Key Environmental Considerations

9.4.1 Soil Health and Conservation

Soil is a non-renewable resource on human timescales and forms the basis of all agricultural productivity. Unsustainable practices can quickly deplete soil nutrients, destroy structure, and lead to erosion and desertification.

Key Soil Degradation Issues:

- **Erosion** – Removal of topsoil by wind or water, exacerbated by deforestation, overgrazing, and poor land management.
- **Nutrient depletion** – Caused by over-reliance on synthetic fertilizers and lack of organic matter replenishment.
- **Soil compaction** – Resulting from the overuse of heavy machinery, reducing root penetration and water infiltration.

Sustainable Soil Management Practices:

- **Conservation tillage** – Reduces disturbance, enhances organic matter retention, and minimizes erosion.
- **Cover cropping** – Plants like legumes protect bare soil, add nitrogen, and support soil biota.
- **Crop rotation** – Prevents pest and disease cycles, improves nutrient cycling, and supports soil structure.
- **Agroforestry** – Combines trees with crops or livestock to improve nutrient recycling, reduce erosion, and enhance microclimates.

Practical Example

In hillside farming regions, terracing can be employed to reduce runoff and prevent soil loss, while windbreaks (lines of trees or shrubs) protect against wind erosion in flat, arid zones.

9.4.2 Water Management and Conservation

Water is essential for all agricultural processes but is increasingly under stress due to overuse, pollution, and climate variability. Efficient water management is crucial for both productivity and environmental sustainability.

Water Management Challenges:

- **Over-extraction** of groundwater leading to aquifer depletion.
- **Contamination** from agricultural runoff containing nitrates, phosphates, and pesticides.
- **Inefficient irrigation** systems that waste water and contribute to salinization.

Sustainable Water Use Practices:

- Drip irrigation systems deliver water directly to the root zone, minimizing evaporation.
- Rainwater harvesting collects and stores rainfall for later use, reducing dependency on external water sources.
- Planting drought-tolerant crops suited to local conditions, reducing overall water demand.
- Mulching and soil cover practices conserve moisture, reduce evaporation, and moderate soil temperature.

Practical Example:

In semi-arid regions, **xeriscaping** with drought-resistant native plants can reduce irrigation needs while maintaining agricultural productivity.

9.4.3 Integrated Pest Management (IPM)

The widespread use of synthetic pesticides has led to serious environmental issues including pesticide resistance, pollution of water bodies, and declines in beneficial insect populations such as pollinators.

Integrated Pest Management (IPM) offers an ecologically sound alternative by combining multiple pest control methods based on scientific monitoring and threshold levels.

Key IPM Strategies:

- **Biological control** – Introducing or conserving natural predators (e.g., parasitic wasps, ladybugs).
- **Cultural practices** – Adjusting planting dates, crop spacing, or using pest-resistant crop varieties.
- **Mechanical controls** – Traps, barriers, and manual removal methods to limit pest access.
- **Selective chemical use** – Only applying pesticides when necessary, and choosing low-toxicity, targeted formulations.

Practical Example:

In vegetable farming, planting marigolds as companion plants can repel nematodes, reducing the need for chemical nematicides.

9.4.4 Biodiversity Conservation

Biodiversity is a cornerstone of sustainable agriculture. It enhances ecosystem services such as pollination, pest control, nutrient cycling, and soil fertility, which are vital for long-term agricultural productivity. However, intensive farming methods often result in the destruction of natural habitats, leading to significant biodiversity loss.

Biodiversity Considerations:

- **Habitat loss:** Conversion of forests, grasslands, and wetlands into agricultural land fragments ecosystems and displaces native flora and fauna.
- **Monoculture:** The dominance of single-crop farming reduces genetic variability and resilience against pests and diseases.

- **Chemical pollution:** The use of synthetic fertilizers, herbicides, and pesticides contaminates soil and water, threatening beneficial insects, amphibians, birds, and aquatic life.

Sustainable Biodiversity Practices:

- **Agroecology:** This holistic approach integrates ecological principles into farming by promoting crop diversity, intercropping, polycultures, and integration with livestock. It fosters habitat complexity, which supports a variety of species.
- **Buffer zones:** Vegetated strips around fields, rivers, and wetlands act as ecological transition zones, reducing runoff and offering habitat to beneficial species.
- **Wildlife corridors:** These are strips of habitat that connect isolated patches of natural environments, facilitating gene flow and animal movement, which are essential for species survival.
- **Conservation agriculture:** Practices like reduced tillage, permanent soil cover, and diversified crop rotations reduce disturbance and support soil biodiversity.

Example:

In Southeast Asia, smallholder farmers practicing agroforestry—planting fruit trees alongside vegetables and cereals—have reported increases in pollinator populations, which led to higher crop yields and improved ecosystem stability.

9.4.5 Climate Resilience

Climate change presents one of the most significant threats to agriculture. Rising temperatures, shifting precipitation patterns, increased frequency of extreme weather events, and altered pest and disease dynamics directly impact crop yields and food security.

Climate Resilience Strategies:

- **Climate-smart agriculture (CSA):** CSA encompasses practices that sustainably increase productivity and incomes, adapt and build resilience to climate change, and reduce or remove greenhouse gas emissions. Examples include precision agriculture, conservation tillage, and rotational grazing.
- **Water conservation:** Implementing smart irrigation technologies such as soil moisture sensors, automated systems, and efficient water application (e.g., drip irrigation) helps conserve water resources.
- **Diversification:** Cultivating a variety of crops and incorporating livestock or aquaculture systems helps spread risk and maintain income in the face of climate variability.
- **Agroclimatic zoning:** Selecting crops suited to the specific climatic conditions of a region reduces reliance on irrigation and artificial inputs.

Example:

In sub-Saharan Africa, farmers adopting conservation agriculture with drought-tolerant maize varieties have seen improved yields even in years with below-average rainfall, demonstrating the importance of resilient farming systems.

9.5 Tools for Environmental Sustainability Analysis

To design and implement environmentally sustainable agricultural projects, it is critical to use tools that assess, monitor, and mitigate environmental impacts.

Tool	Purpose
Environmental Impact Assessment (EIA)	A structured process that identifies and evaluates the potential environmental impacts of a proposed project before implementation.

Carbon Footprint Calculator	Estimates total greenhouse gas emissions from farming activities, helping to identify mitigation opportunities.
Water Footprint Assessment	Measures the volume of freshwater used throughout agricultural production, including blue (surface/ground water), green (rainwater), and grey (polluted water).
Soil Health Indicators	Include physical (e.g., texture, compaction), chemical (e.g., pH, nutrient levels), and biological (e.g., microbial activity) measures to assess soil functionality.
Pesticide Risk Assessment	Evaluates the toxicity, persistence, and bioaccumulation of pesticides to ensure safe usage and prevent ecosystem harm.
Ecological Footprint Calculator	Gauges the amount of land and natural resources required to support a farming system, helping to identify unsustainable practices.

These tools assist stakeholders in decision-making and in aligning agricultural development with sustainability goals.

9.6 Policies and Regulations for Environmental Sustainability

Ensuring environmental sustainability in agriculture is not merely a best practice—it is a legal and ethical requirement. Regulatory compliance ensures legitimacy, opens access to funding, reduces environmental liabilities, and fosters community trust. Agricultural projects must navigate a landscape shaped by both international commitments and local legal frameworks, aligning their practices with global sustainability goals.

Key Frameworks and Guidelines

1. **The Convention on Biological Diversity (CBD)**

   Adopted at the 1992 Earth Summit, the CBD is a foundational international agreement focused on three key objectives:

   - Conservation of biological diversity
   - Sustainable use of its components
   - Fair and equitable sharing of benefits from genetic resources

   Implications for Agriculture:

   - Agricultural projects should protect natural habitats, avoid invasive species, and promote traditional crop varieties and indigenous knowledge.
   - Project designs should include biodiversity assessments and integrate conservation areas into farm layouts.

2. **Local Environmental Regulations**

   These are laws, standards, and policies implemented by national and sub-national environmental agencies. Compliance is typically a prerequisite for obtaining environmental clearances or permits.

   Examples include:

   - **Environmental Impact Assessment (EIA)** regulations requiring baseline studies and mitigation plans.
   - **Water Resource Acts** that govern irrigation practices and groundwater usage.
   - **Pesticide Control Laws** to manage chemical use and residue levels.
   - **Protected Species Legislation**, which restricts activities that harm endangered flora or fauna.

   Project Actions:

   - Secure necessary environmental permits before land development.

- Conduct regular audits and maintain environmental compliance documentation.
- Train staff on legal obligations related to waste disposal, water use, and chemical management.

3. **Sustainable Development Goals (SDGs)**

   The 2030 Agenda for Sustainable Development sets global goals for prosperity while protecting the planet. Agriculture intersects with several of these goals:

   - **SDG 6: Clean Water and Sanitation**
     - Encourages efficient irrigation systems (e.g., drip or sprinkler)
     - Promotes the treatment and reuse of agricultural wastewater
     - Reduces runoff pollution through integrated nutrient management
   - **SDG 13: Climate Action**
     - Supports adoption of climate-smart agriculture (e.g., zero tillage, agroforestry)
     - Promotes carbon sequestration through soil management
     - Encourages renewable energy use in farming operations
   - **SDG 15: Life on Land**
     - Focuses on halting deforestation and land degradation
     - Promotes reforestation, organic farming, and agroecological approaches
     - Encourages land-use planning to conserve ecosystems and reduce habitat fragmentation

Alignment Benefits:

- Projects aligned with SDGs may access **green bonds**, **climate adaptation funds**, and **technical assistance** from international donors and development agencies.

Strategic Benefits of Compliance and Alignment

1. **Access to Financing and Incentives**
   - Many international funding bodies (e.g., World Bank, FAO, GEF) require proof of environmental compliance and alignment with sustainability goals.
   - Governments may offer tax breaks, subsidies, or grants for sustainable practices (e.g., organic certification, solar-powered irrigation).
2. **Enhanced Stakeholder Engagement**
   - Community buy-in is more likely when projects respect local ecosystems and legal obligations.
   - Transparency in regulatory compliance builds trust with local authorities, NGOs, and investors.
3. **Risk Reduction**
   - Legal compliance helps avoid fines, project shutdowns, and reputational damage.
   - Environmental management plans reduce long-term costs associated with land degradation or pollution remediation.

Implementation Tip

Develop an Environmental Compliance Checklist during the project planning phase that includes:

- Relevant local and national regulations
- Required permits and licenses
- Environmental standards to be met (air, water, soil quality)
- Reporting and documentation protocols
- Emergency response plans for environmental incidents

Chapter 10: Risk Assessment and Mitigation Strategies

Benefit:

Equips learners with strategies to identify and mitigate risks in volatile agricultural environments. Researchers can use this content to study the effectiveness of resilience-building measures like crop insurance or climate-smart practices.

10.1 Introduction

Agricultural projects, like any other business ventures, are subject to a wide range of risks. These risks, if not managed properly, can threaten the success, profitability, and sustainability of the project. The dynamic nature of agriculture—including dependence on natural conditions, market volatility, and political factors—makes risk management a critical component of any agricultural project plan.

This chapter explores how to identify, assess, and manage the various risks that may arise throughout the lifecycle of an agricultural project. By understanding and preparing for these risks, project managers can minimize their potential impact and enhance the likelihood of project success.

10.2 What is Risk Assessment in Agricultural Projects?

Risk assessment in agricultural projects is a systematic process aimed at identifying, analyzing, and evaluating uncertainties that may hinder the achievement of project goals. These uncertainties—ranging from climatic variability to market fluctuations—can disrupt operations, delay implementation, or cause financial losses if not properly managed.

The process involves:

- Identifying potential threats or vulnerabilities (e.g., pest infestations, input shortages)
- Estimating the likelihood of these risks occurring
- Assessing the severity of their potential impacts on key project objectives
- Formulating risk mitigation or response strategies

Risk assessment is not a one-time activity but an ongoing, adaptive process integrated into project planning, execution, and monitoring. It equips project managers with foresight and preparedness, enabling informed decision-making, resource optimization, and timely corrective actions.

Benefits of Risk Assessment:

- Prevents project delays and cost overruns
- Enhances stakeholder confidence and investor trust
- Improves operational efficiency and long-term project viability
- Enables strategic planning and contingency preparedness

10.3 Objectives of Risk Assessment and Mitigation

The primary goal of risk assessment and mitigation in agricultural projects is to safeguard the project's objectives and enhance its resilience. The specific objectives include:

- **Identify Risks and Hazards:** Recognize both internal and external threats—natural, economic, operational, and social—that may impact the project.
- **Evaluate Risk Probability and Impact:** Prioritize risks by determining how likely they are to occur and the extent of damage or disruption they could cause.

- **Develop Mitigation and Transfer Strategies:** Implement measures to reduce risk exposure or shift risks to third parties (e.g., through insurance or partnerships).
- **Establish Contingency Plans:** Prepare response strategies for risks that cannot be entirely avoided, ensuring the project remains on track during emergencies.
- **Create Monitoring and Review Mechanisms:** Continuously observe risk indicators and adjust strategies in real time to reflect changing conditions.
- **Strengthen Project Resilience:** Build the capacity of the project and its stakeholders to withstand, recover from, and adapt to adverse conditions.

These objectives support sustainable and adaptive agricultural development, especially in regions prone to instability or environmental stress.

10.4 Types of Risks in Agricultural Projects

Agricultural projects face a broad spectrum of risks that can be grouped into five primary categories: **natural, economic, social, technological, and political**. Understanding these categories helps in designing targeted mitigation and adaptation strategies.

10.4.1 Natural Risks

Natural risks are among the most significant threats to agricultural projects due to their direct impact on crops, livestock, and natural resources. These risks are often beyond human control but can be anticipated and managed with strategic planning.

Examples of Natural Risks:

- **Climate Change and Weather Variability:**

Shifts in seasonal patterns, extreme weather events (such as droughts, floods, storms), and rising temperatures can disrupt planting schedules, damage crops, and reduce yields.

- **Pests and Diseases:**

  Sudden infestations by locusts, fungi, or crop-specific diseases can lead to severe yield losses and increased pesticide usage, harming both production and ecosystems.
- **Soil Degradation:**

  Continuous cropping without replenishing soil nutrients can lead to erosion, salinization, compaction, and decreased fertility, reducing land productivity.
- **Water Scarcity:**

  Over-extraction, droughts, and inefficient irrigation can limit water availability, directly impacting crop performance and livestock health.

Mitigation Strategies for Natural Risks:

- **Climate-Smart Agriculture (CSA):**

  Adopt adaptive practices like conservation agriculture, drought-resistant seeds, no-till farming, water harvesting, and mulching to improve productivity under changing climatic conditions.
- **Integrated Pest Management (IPM):**

  Use a combination of biological control (e.g., natural predators), crop rotation, habitat management, and selective pesticide use to reduce the risk of infestations.
- **Early Warning and Monitoring Systems:**

Leverage technology such as satellite imaging, weather forecasting tools, and pest surveillance platforms to anticipate and prepare for potential hazards.

- **Agroecological Approaches:**
  Promote biodiversity in farming systems (e.g., polycultures, agroforestry) to naturally buffer against environmental shocks and pest outbreaks.

10.4.2 Economic Risks

Economic risks arise from external market dynamics and internal financial management challenges. These risks affect the project's profitability, resource allocation, and long-term sustainability.

Examples of Economic Risks:

- **Market Price Volatility:**
  Agricultural commodity prices are highly sensitive to global supply and demand, trade policies, climate events, and geopolitical tensions, often leading to income instability.
- **Rising Input Costs:**
  The prices of critical inputs such as fertilizers, agrochemicals, labor, and fuel can rise unexpectedly, straining the project budget and affecting profitability.
- **Access to Finance:**
  Limited access to affordable credit, high-interest rates, or stringent loan conditions can constrain working capital, delay investments, and restrict growth.
- **Currency Fluctuations:**
  In export-oriented agricultural ventures, currency depreciation or inflation may affect profitability, especially where inputs are imported.

Mitigation Strategies for Economic Risks:

- **Diversification of Income Streams:**

Avoid overreliance on a single crop or product. Consider intercropping, livestock integration, agrotourism, or value-added processing to stabilize income.

- **Forward Contracts and Futures Trading:**
  Lock in prices with buyers through contractual agreements or use commodity futures markets to hedge against price fluctuations.
- **Efficient Cost Management:**
  Implement precision agriculture, mechanization, and optimized input usage to lower production costs and improve margins.
- **Engage with Subsidy and Credit Schemes:**
  Access government-supported agriculture financing programs, insurance schemes, or partnerships with microfinance institutions and agricultural banks.

10.4.3 Social Risks

Agricultural projects are closely linked to the social fabric of the communities in which they operate. These projects can generate both positive and negative social impacts. Failure to adequately consider the local context and social dynamics may lead to serious setbacks.

Examples of Social Risks

- **Land Conflicts**: In many rural areas, formal land titles are lacking, leading to overlapping claims and disputes. Projects that proceed without resolving such issues risk being halted by legal action or community opposition.
- **Community Resistance**: If local populations perceive the project as a threat—whether due to environmental degradation, loss of livelihood, or exclusion from benefits—they may resist the project through protests, legal challenges, or even sabotage.

- **Labor Issues**: Unskilled or underpaid labor, unsafe working conditions, or the absence of legal contracts can lead to workforce dissatisfaction, low productivity, and in severe cases, strikes or attrition.

Mitigation Strategies for Social Risks

- **Stakeholder Engagement**: Conduct early and continuous consultations with all affected parties—farmers, local leaders, women's groups, indigenous populations—to build trust, gain social license to operate, and incorporate local knowledge into project planning.
- **Clear Land Ownership Documentation**: Partner with legal and land survey institutions to validate land tenure and ownership. Transparent contracts and agreements can reduce future disputes.
- **Fair Labor Practices**: Adopt ethical labor standards by providing competitive wages, fair working hours, social protections, training opportunities, and grievance redress mechanisms.

10.4.4 Technological Risks

Technology plays an increasingly vital role in agricultural productivity, but its adoption also introduces new forms of risk, especially in resource-limited settings where access to training or repair services may be constrained.

Examples of Technological Risks

- **Equipment Failure**: Breakdowns in key equipment such as irrigation pumps, tractors, or processing units can interrupt operations during critical stages like planting or harvesting.
- **Technological Obsolescence**: Rapid changes in agri-tech—such as AI-based decision tools, GPS-guided machinery, or automated sensors—can render existing systems outdated and less competitive.
- **Data Security Risks**: With the growing use of digital platforms for farm management, precision agriculture, and market analytics, there is a risk of cyber threats, data theft, or loss of critical information.

Mitigation Strategies for Technological Risks

- **Routine Maintenance**: Establish a preventive maintenance schedule, ensure spare part availability, and train staff in basic repair and troubleshooting.
- **Technology Upgrades**: Plan for phased adoption of new technology, choosing scalable, user-friendly, and modular systems that can adapt over time.
- **Cybersecurity Measures**: Use secure networks, regularly update software, create access protocols, and maintain data backups to ensure business continuity.

10.4.5 Political and Regulatory Risks

Government policies, legal frameworks, and the broader political environment play a crucial role in shaping the agricultural sector. Political instability or regulatory changes can significantly alter project feasibility.

Examples of Political and Regulatory Risks

- **Policy Changes**: Sudden changes in agricultural subsidies, export bans, or price controls can alter input costs, market access, or revenue projections.
- **Regulatory Compliance**: Complex or inconsistent enforcement of labor laws, environmental rules, or licensing requirements may lead to fines or shutdowns.
- **Land Tenure Laws**: Government expropriation or land redistribution programs, particularly in developing regions, may create uncertainty around long-term land use rights.

Mitigation Strategies for Political Risks

- **Policy Monitoring**: Stay abreast of relevant policy developments and participate in agricultural networks or industry groups to advocate for fair regulations.

- **Regulatory Compliance**: Hire local legal consultants to navigate regulatory requirements and maintain a strong compliance culture across project operations.
- **Political Risk Insurance**: In politically unstable regions, specialized insurance products can protect against expropriation, currency inconvertibility, and civil unrest.

10.5 Risk Mitigation Framework

To effectively manage risk in agricultural projects, a structured approach is required. A robust risk management framework consists of four key stages:

1. **Risk Identification**: Use participatory approaches, expert interviews, field surveys, and tools like SWOT (Strengths, Weaknesses, Opportunities, Threats) and PESTLE (Political, Economic, Social, Technological, Legal, Environmental) analysis to identify all potential threats.
2. **Risk Assessment**: Analyze the likelihood (e.g., rare, possible, likely) and **impact** (e.g., low, moderate, severe) of each risk. This can be visualized using a **risk matrix**, helping prioritize the most critical threats.
3. **Risk Mitigation Planning**: Develop tailored strategies for the top-ranked risks, aiming to reduce their probability or impact. These may include prevention, transfer (e.g., insurance), acceptance (if impact is low), or contingency planning.
4. **Monitoring and Review**: Establish performance indicators and feedback mechanisms to track how risks evolve over time. Periodically update the risk register and mitigation strategies based on new information or changes in the external environment.

10.6 Tools for Risk Assessment and Mitigation

Tool	Purpose
Risk Register	A dynamic log that captures each risk, its description, likelihood, impact, mitigation plan, and responsible party. Essential for ongoing project management.
Risk Matrix	A 2D grid that plots risks based on likelihood and impact, making it easier to prioritize and visualize overall risk exposure.
SWOT Analysis	A strategic tool to evaluate internal strengths/weaknesses and external opportunities/threats that could affect project outcomes.
Sensitivity Analysis	A modeling tool that evaluates how variations in key inputs (e.g., rainfall, prices, labor costs) affect financial or production outcomes. Useful for stress testing scenarios.

10.7 Summary

Risk is an inherent part of agricultural project management. However, with careful planning, stakeholder engagement, and strategic interventions, many of these risks can be anticipated and managed effectively. This chapter provided an in-depth view of the various types of risks—natural, economic, social, technological, and political—that agricultural projects typically face. It also outlined a systematic framework and practical tools for mitigating these risks.

By embedding risk assessment and mitigation in the core of project design and operations, managers can enhance project resilience, attract investment, and ensure the long-term sustainability of agricultural ventures.

In the next chapter, we turn to Monitoring and Evaluation (M&E)—a critical process that allows project teams to track performance, assess whether risk mitigation strategies are effective, and ensure that goals and impacts remain on target.

Chapter 11: Monitoring and Evaluation (M&E)

Benefit:

Students learn to track project progress and outcomes using real-world tools and indicators. Researchers can apply M&E frameworks to assess the effectiveness of pilot projects or development programs.

11.1 Introduction

Monitoring and Evaluation (M&E) are critical elements in the successful management of agricultural projects. They provide the tools and frameworks necessary to assess the progress of a project, ensure that objectives are being met, and determine whether the project is delivering the expected outcomes. M&E systems also help identify challenges, guide decision-making, and ensure that resources are being used efficiently.

This chapter explores the key concepts, tools, and methods involved in monitoring (tracking ongoing activities and outputs) and evaluation (assessing the impact and outcomes of a project) within agricultural projects. By implementing effective M&E practices, project managers can ensure that their agricultural ventures remain on track and adaptable to changing circumstances.

11.2 What is Monitoring and Evaluation?

Monitoring and Evaluation (M&E) is a critical component of agricultural project management that ensures transparency, effectiveness, and continuous improvement throughout the lifecycle of a project.

Monitoring is defined as the *systematic and ongoing* collection, analysis, and use of information to track the progress of project implementation. It involves routine measurement of inputs, activities, outputs, and immediate outcomes. Monitoring helps ensure that the project is proceeding as planned, resources are being used efficiently, and short-term objectives are being met. It acts as a real-time feedback mechanism that enables proactive decision-making.

Evaluation, in contrast, is a more *episodic and reflective* process that assesses the project's achievements, effectiveness, efficiency, relevance, and sustainability. Evaluations are typically conducted at key milestones—such as mid-term or end-of-project stages—and involve a deeper analysis of whether the project has delivered its intended results, the impact it has had on beneficiaries, and what lessons can be drawn for future initiatives.

Together, **Monitoring and Evaluation (M&E)** function as complementary tools that provide a comprehensive understanding of a project's performance. While monitoring focuses on *how the project is being implemented*, evaluation is concerned with *what the project has achieved* and *why* those results occurred. This dual perspective supports evidence-based decision-making, strategic learning, and accountability to stakeholders.

11.3 Objectives of Monitoring and Evaluation

The implementation of a robust M&E framework in agricultural projects serves several interrelated objectives:

Tracking Progress

M&E systems help in systematically tracking the implementation of project activities against predefined plans, timelines, and budgets. Regular monitoring enables managers to verify whether milestones are being met and ensures alignment with project objectives. It answers questions such as:

- Are inputs being delivered on time?
- Are activities being carried out as scheduled?
- Are outputs consistent with expectations?

Assessing Impact

Evaluation plays a key role in determining the broader effects and long-term changes brought about by the project. This includes measuring:

- Increases in agricultural productivity or crop yields
- Improvements in household income or food security
- Positive changes in environmental practices or sustainability
- Enhanced capacity of farmers or agricultural institutions

This objective helps determine the effectiveness and value of the project interventions.

Identifying Problems Early

Continuous monitoring enables early detection of deviations, bottlenecks, or unforeseen issues. By identifying problems as they arise, project managers can implement timely corrective actions, reallocate resources, or modify strategies before they result in major setbacks. This enhances the resilience and responsiveness of the project.

Improving Accountability

M&E provides objective and verifiable data that can be used to demonstrate progress and outcomes to stakeholders, including beneficiaries, funding agencies, government bodies, and other partners. Clear reporting of performance indicators promotes **transparency**, **credibility**, and **trust**, and ensures that all parties are held accountable for their roles and responsibilities.

Learning and Adaptation

One of the most strategic benefits of M&E is the generation of knowledge and learning. By systematically capturing what works, what doesn't, and why, project teams can refine their approaches, replicate successful models, and avoid past mistakes. This **adaptive management** approach fosters continuous improvement and strengthens the design of future agricultural initiatives.

11.4 Key Components of Monitoring and Evaluation

A robust Monitoring and Evaluation (M&E) system is essential for ensuring that agricultural projects are implemented effectively, deliver measurable results, and contribute to long-term development goals. M&E systems are typically divided into two main components: Monitoring and Evaluation, each with its own methodologies, tools, and objectives.

11.4.1 Monitoring

Monitoring refers to the continuous process of systematically collecting, analyzing, and using data to assess how well a project is progressing against its planned objectives. It allows managers to make informed, real-time decisions and to ensure that resources are being used effectively.

Key Components of Monitoring

- **Input Monitoring**

  Focuses on tracking the quantity, quality, and timely delivery of project inputs such as:

  - Financial resources (budget allocations and expenditures)
  - Human resources (staff, labor, technical support)

- Physical resources (land, seeds, fertilizers, irrigation systems, tools) This component ensures that resources are used efficiently and that procurement and distribution align with the project timeline.

- **Process Monitoring**

  Involves assessing the implementation of project activities. It examines whether tasks are being carried out as planned and identifies any procedural bottlenecks or delays. For example:
  - Adherence to planting or harvesting schedules
  - Functioning of irrigation systems
  - Deployment and management of labor This helps in understanding how project activities are translating into outputs and ensures procedural compliance.
- **Output Monitoring**

  Measures the direct and immediate results of project activities. It includes quantifiable achievements such as:
  - Number of hectares cultivated
  - Quantity of crops harvested
  - Livestock raised
  - Infrastructure built (e.g., storage facilities, wells) Outputs serve as early indicators of whether the project is on track toward achieving its outcomes.

Monitoring Tools

- **Progress Reports**

  Regular narrative and financial updates provided by field teams or implementing partners to document activities, challenges, and achievements.

- **Dashboards**

  Interactive digital platforms that visualize key performance indicators (KPIs), allowing for real-time tracking and data-driven decision-making.
- **Site Visits**

  Physical inspections conducted by project managers, evaluators, or stakeholders to observe on-ground implementation and validate reported data.
- **Data Collection Tools**

  Instruments such as structured surveys, mobile apps, GPS trackers, drones, and IoT-based sensors used to gather accurate and timely data from the field.

11.4.2 Evaluation

Evaluation is the structured assessment of the relevance, efficiency, effectiveness, impact, and sustainability of a project. Unlike monitoring, which is ongoing, evaluation is typically conducted at specific intervals—such as mid-term, end-of-project, or post-project—to provide strategic insights into the project's overall success and value.

Key Components of Evaluation

- **Outcome Evaluation**

  Focuses on the intermediate effects of the project on beneficiaries, measuring whether the project has led to desired changes such as:
  - Increased food security
  - Improved household income
  - Enhanced knowledge of sustainable agricultural practices This evaluation helps determine the effectiveness of the project in achieving its objectives.
- **Impact Evaluation**

Examines the long-term and often unintended changes brought about by the project. It answers questions such as:

  - Has the project contributed to rural development?
  - What has been the environmental impact?
  - Has community resilience improved? Impact evaluations provide insights into the broader transformational effects of the intervention.

- **Cost-Benefit Analysis (CBA)**

  Assesses the economic efficiency of the project by comparing the monetary value of benefits (e.g., yield gains, income increases) to the costs incurred (e.g., labor, materials, capital investment). CBA helps determine whether the project was a financially sound investment.

- **Sustainability Evaluation**

  Investigates the extent to which the project's benefits will continue after external support ends. It looks at factors such as:

  - Institutional capacity
  - Community ownership
  - Availability of ongoing financial or technical support Sustainability evaluation is vital for understanding the durability of project outcomes.

Evaluation Tools

- **Surveys and Interviews**

  Structured or semi-structured instruments used to gather firsthand feedback from beneficiaries, stakeholders, and project staff.

- **Focus Group Discussions (FGDs)**

  Facilitated group sessions that provide qualitative insights into community perceptions, behavioral changes, and unintended effects.

- **Case Studies**

In-depth documentation and analysis of specific communities, households, or individuals to illustrate project impacts and contextual factors.

- **Statistical Analysis**

  Use of quantitative methods to analyze project performance metrics, assess trends over time, and compare results against baseline or control data.

11.5 Developing a Monitoring and Evaluation Framework

A robust Monitoring and Evaluation (M&E) framework serves as the backbone for assessing the progress, impact, and sustainability of agricultural projects. It ensures that project activities align with the intended goals and enables project managers to adjust strategies based on evidence. By defining clear objectives, selecting appropriate indicators, and using systematic data collection and reporting methods, the framework helps ensure transparency, accountability, and continuous improvement.

Key Steps in Developing an M&E Framework

1. Define Clear Project Goals and Objectives

The foundation of a successful M&E framework lies in establishing SMART goals—specific, measurable, achievable, relevant, and time-bound objectives that guide the project's direction. These goals should be aligned with the broader vision of the agricultural initiative and provide a clear roadmap for monitoring progress. For example:

- Specific: Increase maize yield.
- Measurable: By 20%.
- Achievable: Within three years.
- Relevant: The goal addresses food security challenges in the region.
- Time-bound: Completion within a 3-year timeframe.

Having well-defined objectives allows stakeholders to have a common understanding of what success looks like, and provides a concrete framework for evaluating progress.

2. Establish Indicators

Indicators are the quantitative and qualitative measures used to track progress towards achieving the project's objectives. These should be:

- Directly linked to the objectives: For example, if the goal is to increase crop yield, an appropriate indicator would be the "yield per hectare" or "total crop output."
- Sufficiently sensitive to detect any changes in the project area.
- Feasible and cost-effective: Consider the availability of resources to measure the indicator accurately.

Types of Indicators:

- **Quantitative Indicators**: These provide measurable, objective data (e.g., crop yield per hectare, income per farmer, number of hectares under sustainable management).
- **Qualitative Indicators**: These measure subjective experiences or perceptions (e.g., farmer satisfaction, community engagement, local knowledge transfer).

Common agricultural indicators include:

- Yield per hectare (for crops or livestock)
- Water use efficiency
- Income per unit of input
- Changes in soil health
- Biodiversity indices

These indicators help provide a clear picture of the project's success and areas needing attention.

3. Develop a Data Collection Plan

A **data collection plan** outlines how the data will be gathered, analyzed, and reported to track project performance and assess outcomes. This plan should include:

- **Types of Data**: Define what specific data will be collected, such as crop productivity, environmental indicators (e.g., soil health), or socioeconomic impact (e.g., income increases for farmers).
- **Methods of Collection**: Choose the most appropriate methods for gathering data, such as:
  - **Surveys**: To collect data from beneficiaries and stakeholders.
  - **Remote Sensing**: To monitor land use, crop health, and water resources.
  - **Field Observations**: For direct tracking of farming practices and outcomes.
  - **Sensors and IoT Tools**: For real-time data on environmental factors like soil moisture or temperature.
- **Frequency of Data Collection**: Decide how often data will be collected—this could be monthly, quarterly, or annually depending on the project's timeline and data requirements.

Ensure that the plan accounts for seasonality, availability of resources, and the need for real-time or longitudinal data.

4. Assign Responsibilities

Clearly defined roles and responsibilities ensure that each aspect of monitoring and evaluation is handled by the appropriate personnel. This includes:

- **Project Manager**: Oversees the M&E process and ensures that data collection methods are followed correctly.
- **Field Officers/Researchers**: Collect primary data, provide insights from the field, and report any anomalies or trends.

- **Data Analysts**: Analyze collected data, generate reports, and interpret findings for decision-making.
- **Stakeholder Involvement**: Engage local communities, farmers, government agencies, or NGOs to support data collection or to provide feedback on project impacts.

Establishing a clear reporting structure ensures that the flow of information is effective and efficient, allowing for prompt adjustments as necessary.

5. Reporting and Feedback Mechanisms

A comprehensive **reporting system** ensures that the results of monitoring and evaluation activities are communicated to relevant stakeholders in a timely and transparent manner. This involves:

- **Regular Progress Reports**: These documents track activities, outputs, challenges, and success stories. They should be accessible to project stakeholders, including funding agencies and local partners.
- **Dashboards**: Interactive, digital platforms that visually display key performance indicators (KPIs) in real-time, allowing managers to monitor progress at a glance.
- **Annual Reviews**: A comprehensive review of the project's progress, challenges, and outcomes over the year. This provides an opportunity for reflection and strategic adjustments.
- **Stakeholder Feedback**: Collect feedback from all relevant parties, including farmers, local communities, and government bodies, to continuously refine project strategies.

A feedback loop ensures that the project adapts to changing circumstances, and lessons learned are integrated into future planning.

11.6 Key Performance Indicators (KPIs) for Agricultural Projects

KPIs are crucial metrics that reflect the performance and success of agricultural projects. These indicators help project managers gauge whether objectives are being met and identify areas for improvement. Some common KPIs include:

- **Yield per Hectare**: Measures productivity and efficiency in crop or livestock production.
- **Revenue per Unit of Input**: Tracks financial returns relative to investments in resources (e.g., seeds, labor, irrigation).
- **Water Use Efficiency**: Measures how effectively water resources are being managed, important for areas facing water scarcity.
- **Soil Health Indicators**: Includes parameters like soil organic matter content, pH levels, and nutrient availability, which are essential for sustainable agriculture.
- **Income Increases for Beneficiaries**: Evaluates the socioeconomic impact of the project on local farmers or communities.
- **Biodiversity Indices**: Assesses the variety and health of plant and animal species within the project area, which is critical for ecological balance.

11.7 Challenges in Monitoring and Evaluation

While M&E systems are essential, several challenges can hinder their effectiveness, including:

- **Data Collection Issues**: Inconsistent methods, lack of skilled personnel, or inadequate tools can lead to inaccurate data, undermining decision-making.
- **Funding Constraints**: A lack of budget for M&E activities may limit the quality, scope, or frequency of monitoring efforts.
- **Stakeholder Engagement**: It can be challenging to get consistent input from all relevant stakeholders, especially in remote or rural areas.

- **Contextual Changes**: Changes in environmental, economic, or political conditions may affect the relevance of the project goals and the data collected.

To overcome these challenges, it is critical to establish a flexible, adaptive M&E system that can accommodate changes and engage stakeholders throughout the project's lifecycle.

11.8 Summary

In agricultural project management, an effective M&E framework is vital for ensuring the project meets its goals, remains on track, and produces meaningful, long-term impacts. By defining clear objectives, selecting appropriate indicators, developing comprehensive data collection plans, and establishing transparent reporting mechanisms, agricultural projects can continuously monitor their progress and make informed decisions to adapt and improve. In the next chapter, we will explore Lessons Learned from Case Studies and Best Practices from successful agricultural projects.

Part IV: Practical Applications and Case Studies

Focus: Real-world practices, success stories, and strategic lessons.

Chapter 12: Case Studies and Best Practices

Benefits:

Chapter 12 provides students and researchers with a rich collection of best practices across various agricultural projects. It emphasizes strategic planning, resource optimization, and stakeholder collaboration. This knowledge base supports evidence-based decision-making and inspires innovative approaches for enhancing project sustainability and long-term success.

12.1 Introduction

Case studies and best practices provide invaluable insights into the real-world application of agricultural project management principles. By examining successful projects, we can identify strategies that work, learn from past mistakes, and apply those lessons to new initiatives. This chapter presents case studies from various agricultural sectors, including smallholder farmer cooperatives, large agribusiness ventures, and sustainable farming initiatives.

These case studies illustrate diverse project types, from crop production to livestock farming, and demonstrate the application of feasibility studies, project planning, and risk management. The aim is to provide practical examples and strategies that can be adopted or adapted for future agricultural projects.

12.2 Case Study 1: Smallholder Farmer Cooperatives

Project Overview

In this case, a group of smallholder farmers in East Africa formed a cooperative with the goal of improving their market access and increasing the profitability of their crops. The cooperative primarily focused on maize, soybeans, and vegetables, aiming to create a more reliable supply chain to local markets while also enhancing the financial stability of the farmers involved.

The primary goals of the cooperative were:

- Improving crop yields by introducing better farming practices.
- Reducing post-harvest losses through better storage facilities.
- Increasing income by ensuring better market access and fair pricing for the farmers' produce.

Key Challenges Faced

Several key challenges were identified at the start of the project, all of which had a direct impact on the farmers' ability to be profitable and sustainable:

1. **Limited Access to Markets:**
    - Farmers were selling their produce at low prices due to a lack of organized market access. They often faced a sellers' market, where they had to sell at whatever price was available, often below their production costs.
2. **Inadequate Infrastructure:**
    - The region suffered from poor road networks, which made it difficult for farmers to transport their produce to markets in a timely manner. This lack of storage facilities and unreliable transportation systems caused crops to spoil, leading to significant post-harvest losses and lost income.
3. **Lack of Technical Knowledge:**
    - Many of the farmers lacked the necessary training in modern farming techniques. As a result, crop yields were suboptimal, and the farmers weren't taking advantage of advanced agricultural practices like soil fertility management, pest control, and proper post-harvest handling.

Project Implementation

To address these challenges, the cooperative employed a variety of strategies designed to create more sustainable farming practices, reduce inefficiencies, and ensure better market access.

1. **Market Linkages:**
   - The cooperative formed partnerships with local agribusinesses and retailers, which guaranteed a fair price for the produce and regular market access. This relationship ensured that the farmers had steady buyers, reducing the unpredictability of the market and the pressure to sell at low prices.
   - These market linkages were pivotal in improving pricing stability, and ensured that the farmers were able to sell at higher prices, thus increasing their income.
2. **Capacity Building:**
   - Training programs were organized for the farmers to help them learn modern farming techniques. These included:
     - Improved farming techniques such as crop rotation, irrigation management, and pest control.
     - Soil fertility management to improve soil health and increase yields over time.
     - Post-harvest handling to reduce crop losses due to poor storage conditions or improper harvesting techniques.
   - Capacity building was designed to equip farmers with the skills and knowledge necessary to improve their productivity and to ensure that they were always up-to-date with current best practices.
3. **Infrastructure Development:**
   - The cooperative facilitated the construction of community-owned storage facilities to preserve crops for longer periods, reducing post-harvest losses.

- The cooperative also worked closely with local governments to improve road access, ensuring that farmers could get their produce to market faster and at lower transportation costs.

4. **Risk Pooling:**
    - The cooperative pooled resources for collective purchasing of inputs, such as fertilizers, seeds, and equipment. This strategy helped to reduce input costs for individual farmers, as bulk purchasing allowed the cooperative to benefit from economies of scale.
    - Risk pooling also provided farmers with a safety net, ensuring that they were less vulnerable to price fluctuations or market shocks.

Outcomes

The strategies implemented by the cooperative led to several positive outcomes that greatly benefited the farmers involved:

1. **Increased Yields:**
    - Maize yields increased by 30% in the first year of the project.
    - Soybean yields rose by 25% in the same period.

    This was a direct result of the improved farming practices and capacity-building efforts, along with better soil management and pest control techniques.

2. **Improved Market Access:**
    - By establishing market linkages with local agribusinesses and retailers, the cooperative was able to secure better prices for their produce. Farmers no longer had to rely on middlemen, who often offered low prices.
    - As a result, the farmers were able to sell their produce at fairer prices, improving their income and creating a more stable source of revenue.

3. **Reduced Post-Harvest Losses:**
    - The construction of community-owned storage facilities and the improvement of transportation infrastructure led to a reduction in post-harvest losses, which had been a significant challenge for the farmers in the past.
    - Farmers could store their crops in optimal conditions, preserving their quality and enabling them to sell at the right time for better prices.
4. **Economic Security:**
    - The cooperative model provided the farmers with a sense of economic security by giving them bargaining power when negotiating with buyers and input suppliers.
    - Additionally, pooling resources allowed the cooperative to better manage financial risks and improve overall productivity.

Lessons Learned

The smallholder farmer cooperative case offers several valuable lessons for agricultural projects aiming to support smallholder farmers and improve their livelihoods:

1. **Collaboration and Collective Action:**
    - One of the most powerful lessons from this case is that collaboration and collective action can significantly improve the resilience and success of smallholder farmers. By working together in a cooperative model, farmers gained access to resources, markets, and training that would have been unavailable to them individually.
    - The cooperative approach not only increased market access but also enhanced bargaining power, allowing farmers to secure better prices for their produce and reduce costs through collective purchasing.

2. **Importance of Infrastructure and Market Access:**
   - Infrastructure development was a critical factor in the success of the project. Improving storage facilities and road access enabled farmers to overcome challenges related to transportation and post-harvest losses.
   - Ensuring that market access is part of the project design is crucial for the economic viability of smallholder farming. Without a stable market, even the best farming practices may not lead to profitability.
3. **Ongoing Capacity-Building:**
   - Capacity-building is not a one-time effort but should be an ongoing process. The training programs offered to farmers should be regularly updated to keep them informed about the latest agricultural technologies, market trends, and environmental practices.
   - This ensures that farmers are equipped with the latest knowledge and skills to adapt to evolving challenges in agriculture, such as changing climate conditions or market demands.

Conclusion

This case study demonstrates that smallholder farmer cooperatives can overcome significant barriers to profitability and sustainability when they focus on market access, infrastructure development, and capacity building. By leveraging collective action, the farmers were able to increase their yields, reduce post-harvest losses, and improve their economic conditions. The cooperative model provided a platform for farmers to gain bargaining power and secure fair prices for their crops, ultimately leading to long-term financial stability and success.

This case illustrates the transformative power of collaboration, strategic partnerships, and sustainable agricultural practices for smallholder farmers in developing regions.

12.3 Case Study 2: Commercial Agribusiness Ventures

Project Overview

A large-scale agribusiness in Brazil invested in a commercial farming venture focused on producing soybeans and corn for export to global markets. The project was part of a broader effort to address global food security by increasing production capacity while maintaining environmental sustainability.

Key Challenges Faced

- **Environmental concerns**: There were concerns about the impact of large-scale commercial farming on deforestation and biodiversity.
- **Labor force management**: The project required a significant number of seasonal laborers, leading to concerns about working conditions and labor rights.
- **Market competition**: The agribusiness faced competition from other countries with lower production costs.

Project Implementation

1. **Sustainable Farming Practices:** The project implemented precision agriculture techniques, including drip irrigation, drought-resistant seed varieties, and crop rotation to minimize environmental degradation and increase productivity.
2. **Certification Programs:** The project invested in obtaining sustainability certifications such as Fair Trade and Rainforest Alliance to enhance brand reputation and appeal to international markets.
3. **Labor Welfare**: The project adhered to international labor standards, offering fair wages, safe working conditions, and benefits to seasonal workers.
4. **Technology Integration**: The agribusiness adopted cutting-edge technologies like drones for monitoring crop health and AI-powered forecasting tools to optimize planting and harvesting schedules.

Outcomes

- The project achieved high productivity levels, with yields increasing by 40% over three years.
- The sustainability certifications boosted market access, allowing the company to sell its produce at premium prices.
- Environmental impacts were minimized through eco-friendly farming methods that protected natural resources while maintaining profitability.

Lessons Learned

- Incorporating sustainability into large-scale agribusiness ventures can lead to both environmental and financial success.
- Labor rights should be prioritized to ensure a stable and productive workforce.
- Technological integration can significantly enhance farm management efficiency and yield prediction.

12.4 Case Study 3: Sustainable Farming in Arid Regions

Project Overview

In this case study, a sustainable farming initiative was launched in an arid region in the Middle East. The focus of the project was to cultivate high-value crops such as dates, medicinal herbs, and organic vegetables using limited water resources. Despite the harsh climatic conditions—characterized by severe water scarcity and saline soils—the project aimed to demonstrate that agriculture could not only be profitable but also environmentally sustainable in arid regions.

The project set out to:

- **Produce high-value, drought-resistant crops** that could thrive with minimal water.

- **Showcase sustainable farming techniques** that could be replicated in similar regions around the world.
- **Promote long-term environmental sustainability** while supporting local communities economically.

Key Challenges Faced

The farming initiative in this arid region faced several significant challenges due to the natural environment:

1. **Severe Water Scarcity:**
   - Water is the most critical resource in arid and semi-arid regions, and this project faced significant challenges in ensuring a steady and sufficient supply of water for irrigation. The region had insufficient rainfall and relied heavily on groundwater, which is both limited and often costly to access.
2. **Soil Salinity:**
   - The local soil was naturally saline, meaning that it had a high concentration of salts, which made it unsuitable for growing traditional crops. Salinity can hinder the ability of plants to absorb water and nutrients, which significantly reduces crop yields unless proper management techniques are applied.
3. **High Initial Investment:**
   - To manage these challenges, the project required a significant upfront investment in innovative technologies, including systems for water desalination, advanced irrigation methods, and techniques to restore soil health. These technologies were essential but costly to implement, requiring careful financial planning and support.

Project Implementation

To address these challenges, the project adopted a series of innovative strategies aimed at sustainable agriculture, maximizing the efficient use of water and improving soil health while maintaining profitability.

1. **Water Management Innovations:**
   - Solar-powered desalination plants were introduced to convert saline groundwater into usable freshwater. This technology helped reduce dependence on freshwater sources and provided a more sustainable way to supply water to crops.
   - The project also implemented drip irrigation systems, which deliver water directly to the roots of plants in small, controlled amounts. This system minimizes water loss due to evaporation and runoff, making it more water-efficient compared to traditional irrigation methods.
2. **Soil Management:**
   - To tackle soil salinity, the project used biochar—a form of charcoal produced from organic matter that helps improve soil structure, increase nutrient retention, and reduce the harmful effects of salt. Biochar has been shown to enhance soil fertility and reduce the leaching of salts into groundwater.
   - Additionally, composting techniques were used to improve organic matter content in the soil, enhancing its ability to retain moisture and improve overall soil health. These organic methods were designed to counteract the harmful effects of soil salinity and improve crop productivity over time.
3. **Diversification of Crops:**
   - The project focused on growing high-value crops that are better suited to arid climates and can tolerate low water availability and saline soils. These crops included dates (which are naturally

drought-tolerant), medicinal herbs (such as thyme and lavender), and organic vegetables.

   - Diversification of crops reduced the risk of crop failure, ensuring that the project had multiple revenue streams. This approach also helped the project maintain soil fertility by rotating different crops and avoiding monoculture farming, which can degrade the soil.

4. **Community Engagement:**
   - One of the key aspects of the project was involving **local farmers** in the implementation process. The initiative offered training sessions on water conservation, organic farming practices, and modern irrigation techniques.
   - By training local farmers, the project ensured that the knowledge and skills necessary for sustainable farming practices were passed on to the wider community, helping to empower local populations and ensuring that the project could be sustained beyond its initial phase.

Outcomes

The project achieved several significant outcomes, demonstrating that sustainable farming in arid regions is possible, even with limited water and poor soil conditions.

1. **Successful Crop Production:**
   - The project successfully produced high-quality organic crops despite the challenging environmental conditions. Crops like dates and medicinal herbs thrived, and the vegetables produced were of high market value. The project showed that arid regions can support profitable agricultural ventures if the right technologies and practices are employed.

2. **Reduction in Water Usage:**
    - Water usage was reduced by 50% through the adoption of efficient drip irrigation systems and solar-powered desalination plants. The efficient water management practices allowed the project to grow crops with significantly less water than traditional farming methods, making it more sustainable and less reliant on depleting water resources.
3. **Soil Health Improvement:**
    - The use of biochar and composting techniques led to significant improvements in soil health. Over time, the effects of soil salinity were reduced, and the soil became more fertile, which in turn improved crop yields. The organic farming techniques promoted by the project also helped ensure that the farming practices remained environmentally sustainable.
4. **Job Creation and Economic Benefits:**
    - The project generated **jobs** for local communities, both in farming and in related sectors such as water management and organic produce processing. The initiative also contributed to the **economic development** of the region by creating new sources of income and improving the livelihoods of local farmers.
5. **Environmental Sustainability:**
    - By using solar-powered desalination and organic farming methods, the project demonstrated that agriculture could be environmentally sustainable even in challenging conditions. The use of renewable energy for water desalination and the promotion of soil conservation techniques reduced the ecological footprint of the project.

Lessons Learned

The Sustainable Farming in Arid Regions case provides several valuable insights for future agricultural projects, especially in regions facing similar environmental challenges:

1. **Water Management is Crucial:**
    - Innovative water management technologies, such as solar-powered desalination and efficient drip irrigation systems, are critical for making farming sustainable in water-scarce areas. These systems can help reduce water waste, improve crop yields, and ensure that agriculture remains viable in the long term.
2. **Soil Health Improvements are Key:**
    - Soil salinity is a significant barrier to successful farming in arid regions, but it can be mitigated through the use of biochar and composting. Maintaining and improving soil health is critical for ensuring long-term agricultural success, especially in regions where the soil is naturally saline.
3. **Community Involvement Ensures Long-Term Success:**
    - Engaging local communities in the project is essential for its sustainability. Training local farmers on water conservation, organic farming techniques, and modern irrigation systems ensures that the knowledge gained through the project can be applied beyond the life of the project itself.
4. **Diversification Reduces Risks:**
    - Crop diversification helps spread risk by ensuring that the project is not dependent on a single crop. Growing high-value crops suited to arid climates, like dates and medicinal herbs, allowed the project to remain profitable despite the challenges of limited water and soil salinity.

Conclusion

This case study demonstrates that with the right innovative technologies, sustainable practices, and community engagement, agriculture can thrive even in the most challenging environments, such as arid regions. The Sustainable Farming in Arid Regions project successfully produced high-quality crops while addressing key challenges like water scarcity and soil salinity. By focusing on efficient water management, soil health improvement, and crop diversification, the project was able to ensure both profitability and environmental sustainability, offering a model for future agricultural initiatives in similar regions around the world.

12.5 Best Practices for Agricultural Projects

From the case studies presented in the chapter, several best practices have emerged that can guide future agricultural projects in diverse contexts. These practices are rooted in the principles of sustainability, efficiency, and community involvement. Here's a deeper dive into the best practices that should be considered:

1. Incorporate Sustainability

Environmental sustainability is not just a buzzword but a core component of agricultural projects, especially as climate change and resource depletion become more pressing global issues. Regardless of the scale—whether in large-scale agribusiness or smallholder farming—ensuring sustainability should be central to all agricultural projects. This can be achieved by:

- **Minimizing resource use**: Projects should aim to reduce water and energy consumption through innovations like drip irrigation or solar-powered equipment. In many projects, such as those in arid regions, managing scarce resources efficiently is crucial.

- **Soil health management**: Maintaining and improving soil health through practices like composting, biochar use, and crop rotation is essential to prevent degradation and promote long-term productivity.
- **Biodiversity preservation**: Encouraging crop diversification not only ensures stable income streams but also helps conserve local ecosystems, promoting a more resilient agricultural system.

Sustainability extends beyond just the environment—it also involves economic sustainability, ensuring that farmers have access to markets and technologies that can provide consistent income over time.

2. Leverage Technology

Technology has the potential to significantly improve the efficiency and profitability of agricultural projects. Several technological advancements have been proven to enhance productivity and sustainability:

- **Precision Agriculture**: Using technologies like GPS, satellite imagery, and sensors to monitor and manage field variability allows for more efficient use of inputs like water, fertilizers, and pesticides, reducing waste and costs.
- **Drone Monitoring**: Drones can provide real-time data on crop health, soil conditions, and irrigation needs, enabling farmers to make quick, informed decisions and target interventions where they are most needed.
- **AI-based Forecasting**: Artificial Intelligence can help farmers with weather forecasting, crop prediction, and disease management by analyzing large datasets and providing actionable insights. This can improve planning, reduce crop failure risk, and increase overall farm efficiency.

These technologies not only enhance productivity but can also make farming more environmentally friendly by reducing waste and enabling better resource management.

3. Community Engagement

Agricultural projects, particularly those that involve smallholder farmers, must prioritize community involvement in all stages, from decision-making to implementation. Engaging local communities ensures that the project is aligned with the needs and realities of the people it serves. Key practices include:

- **Involving communities in planning**: Farmers and local stakeholders should be included in the planning phase of the project to ensure that the proposed strategies are practical and meet their needs.
- **Providing training and capacity building**: Ensuring farmers have access to training on best practices, new technologies, and sustainable farming techniques is essential. This will help them adapt to evolving agricultural trends, enhance productivity, and achieve long-term success.
- **Promoting ownership**: When communities have a stake in the project's success, they are more likely to commit to its goals and outcomes. Encouraging collective decision-making and shared responsibility can strengthen community bonds and ensure the project's sustainability.

Community engagement also involves establishing clear communication channels with all relevant stakeholders—farmers, local governments, agribusinesses, and NGOs—to foster collaboration and ensure that each group contributes to the project's success.

4. Diversification and Risk Management

Agricultural projects are inherently risky, due to factors such as climate change, market volatility, and pest/disease outbreaks. To mitigate these risks, diversification and robust risk management strategies are crucial.

- **Diversification of crops**: Growing a variety of crops that can withstand different environmental conditions ensures that farmers have multiple income streams, reducing the impact of market fluctuations or crop failure from one particular crop.

- **Diversification of income sources**: In addition to crops, agricultural projects can consider integrating livestock farming, aquaculture, or value-added products (e.g., processed food, biofuels) into their business models. This can increase overall resilience and stabilize income over the long term.
- **Risk-sharing mechanisms**: Projects can consider insurance schemes, community-based risk pooling, or contract farming to share risks among stakeholders and protect farmers from unforeseen events.

Properly implemented risk management strategies can significantly reduce the vulnerability of farming communities and ensure the long-term sustainability of agricultural projects.

5. Capacity Building

Continuous training and skills development are vital for the success and sustainability of agricultural projects. Farmers need to stay updated on the latest techniques, technologies, and market trends to remain competitive. Capacity building includes:

- **Training on sustainable practices**: Providing training on conservation agriculture, soil fertility management, and integrated pest management helps farmers improve yields and reduce costs.
- **Financial literacy**: Educating farmers on budgeting, financial planning, and market trends can help them make informed decisions and manage their finances better.
- **Technology adoption**: Ensuring farmers are proficient in using new technologies—such as mobile apps for weather forecasting or digital platforms for market access—helps them increase efficiency and profitability.

Ongoing capacity building also enables farmers to adapt to changes, such as shifts in market demand or emerging agricultural trends. It is a powerful tool for improving productivity, reducing vulnerability to risks, and supporting farmers' long-term success.

12.6 Conclusion

The case studies and best practices outlined in this chapter underscore the fundamental principles that contribute to the success of agricultural projects, regardless of their scale or geographical location. Successful projects integrate:

- **Sustainability** into every facet of their design and operation, ensuring that the environment, economy, and society all benefit from the initiative.
- **Technological innovation** to enhance farm productivity, efficiency, and decision-making, thus ensuring that resources are used effectively.
- **Community engagement** to ensure the involvement of local stakeholders, fostering a sense of ownership and ensuring that projects meet the real needs of farmers.
- **Risk management and diversification** to protect farmers from unpredictable factors and ensure stable income streams.
- **Capacity building** to equip farmers with the knowledge and skills required to thrive in an evolving agricultural landscape.

As agricultural projects face growing challenges from climate change, population growth, and economic instability, the best practices discussed will continue to serve as a blueprint for shaping the agricultural landscape in the future. These practices will help guide future initiatives to become more resilient, profitable, and environmentally sustainable, providing long-term solutions for global food security and rural development.

Chapter 13: Future Technologies and Practices in Agricultural Project Management

Benefit:

Keeps students and researchers current with technological and policy developments. Ideal for thesis topics, research proposals, or foresight studies. Encourages innovation in agricultural entrepreneurship or development programming.

13.1 Introduction

Agriculture is one of the oldest and most vital industries in the world, but it is undergoing rapid transformation. With the challenges posed by climate change, technological advances, and evolving market demands, agricultural projects must adapt to stay competitive and sustainable. This chapter explores the future trends in agricultural project management, highlighting innovations, new practices, and strategies that will drive success in the coming decades.

These trends are not just about improving productivity—they also focus on sustainability, efficiency, and resilience, addressing the increasing global demand for food, fiber, and fuel while minimizing environmental impact. Understanding these trends will equip agricultural project managers with the tools needed to navigate an increasingly complex landscape.

13.2 Digital Agriculture: The Rise of Technology

As technology rapidly evolves, digital agriculture is revolutionizing farming practices by integrating advanced tools and techniques to enhance productivity, sustainability, and efficiency. The use of data-driven technologies allows farmers to optimize their operations, reduce waste, and increase yields while minimizing environmental impacts. Below, we explore some of the most prominent technologies transforming agriculture today.

13.2.1 Precision Agriculture

Precision agriculture involves the use of advanced technologies such as drones, satellite imagery, GPS, and IoT sensors to monitor and manage farming activities with high precision. These technologies collect real-time data that allows farmers to make informed, data-driven decisions.

Benefits of Precision Agriculture:

- **Increased Yields:** By optimizing the use of inputs such as water, fertilizers, and pesticides based on real-time data, farmers can significantly increase crop yields. For instance, soil sensors can provide insights into nutrient levels, helping farmers apply the right amount of fertilizers at the right time.
- **Resource Efficiency:** Precision farming minimizes the use of water, fertilizers, and pesticides by ensuring that they are applied only where needed. This results in lower costs and reduced environmental impact, as resources are not overused. For example, drip irrigation powered by IoT sensors only irrigates the areas that need it, conserving water.
- **Targeted Farming Practices:** Rather than applying uniform treatments across an entire field, precision agriculture allows for targeted actions in specific areas. This means that farmers can address variability in soil type, pest pressure, and moisture content in different parts of the field, leading to a more efficient and tailored approach to farming.
- **Real-Time Decision Making:** Precision agriculture allows for immediate response to changes on the ground. For example, a farmer can instantly detect a pest infestation or a disease outbreak in a specific section of their farm and take corrective action promptly.

The tools in precision agriculture allow farmers to become more effective stewards of the land, ensuring that their farming practices are both economically and environmentally sustainable.

13.2.2 Artificial Intelligence (AI) and Machine Learning (ML)

AI and machine learning are revolutionizing agriculture by enabling the automation of tasks, predictive analytics, and improved decision-making processes. By analyzing vast amounts of data, AI can identify patterns and trends that would be nearly impossible for a human to detect.

Applications of AI in Agriculture:

- **AI in Crop Management:** AI technologies can process complex data sets, such as weather patterns, soil health data, and crop performance metrics. Machine learning algorithms can predict crop growth, identify potential challenges, and offer optimized farming solutions. For instance, AI can recommend the best planting times, irrigation schedules, and even crop rotation strategies based on environmental conditions.
- **Predictive Analytics for Risk Management:** AI systems can analyze historical data and predict weather events, pest outbreaks, and disease risks that could potentially threaten crops. This allows farmers to proactively address issues before they escalate. For example, predictive models can suggest pest control measures before a pest infestation becomes severe.
- **Optimizing Supply Chains:** AI can also improve logistics by forecasting supply chain needs, predicting demand, and optimizing transportation routes. This ensures that farmers' produce reaches the market efficiently and reduces food waste.

By leveraging AI and ML, agricultural operations can improve in accuracy, efficiency, and resilience, ultimately leading to higher yields and lower costs.

13.2.3 Blockchain Technology in Agriculture

Blockchain technology provides a secure, transparent method for tracking products along the agricultural supply chain. Its decentralized and immutable nature ensures that data regarding the origin, processing, and distribution of agricultural products is both tamper-proof and transparent.

Benefits of Blockchain in Agriculture:

- **Traceability:** One of the main advantages of blockchain is the ability to trace the journey of food products from farm to consumer. This is particularly valuable for verifying claims such as organic or sustainable certifications. Blockchain helps build consumer trust by ensuring that food products meet regulatory standards and ethical practices.
- **Supply Chain Optimization:** Blockchain enables real-time tracking of products, reducing delays and inefficiencies in the supply chain. It also minimizes the risk of fraud (such as the sale of counterfeit goods) and ensures more transparent payment systems. Farmers and traders can have more security and confidence in the transactions that occur along the supply chain.
- **Food Safety:** Blockchain's ability to store data at every stage of the production and distribution process ensures that if a food safety issue arises (such as contamination), products can be traced back to their source quickly. This reduces the risk of widespread outbreaks and allows for swift corrective action.

The transparency and efficiency offered by blockchain can strengthen agricultural supply chains and provide a higher level of consumer confidence in food products.

13.3 Climate-Smart Agriculture (CSA)

With climate change becoming an increasingly critical issue for agriculture, Climate-Smart Agriculture (CSA) aims to make agricultural systems more resilient to environmental challenges while mitigating the sector's impact on climate change. CSA combines sustainable practices with modern innovations to adapt to changing climatic conditions, increase productivity, and reduce agriculture's greenhouse gas emissions.

13.3.1 Key Strategies in CSA

- **Climate-Resilient Crops:** The development of genetically modified (GM) crops or hybrid varieties that can withstand extreme weather conditions (droughts, floods, heat) is a key focus in CSA. These crops will be more adaptable to unpredictable climates, ensuring consistent food production and securing food systems globally.
- **Sustainable Land Management:** Practices like crop rotation, agroforestry, and no-till farming help preserve soil health, reduce erosion, and improve water retention. By promoting biodiversity and improving soil structure, these practices enhance the resilience of farming systems to climate-related shocks.
- **Water Management Innovations:** Efficient water use is a cornerstone of CSA, particularly in regions where water is scarce. Techniques like drip irrigation, rainwater harvesting, and desalination plants are crucial to ensuring that water resources are used effectively. These technologies help ensure that crops receive adequate water while minimizing waste and promoting sustainability.

13.3.2 Carbon Farming

Carbon farming involves agricultural practices that sequester carbon in the soil, helping to reduce atmospheric CO2 levels and mitigate climate change. These practices not only help combat global warming but can also enhance soil fertility and farm productivity.

Key Practices in Carbon Farming:

- **Cover Cropping:** Planting cover crops during the off-season helps prevent soil erosion, enhances soil structure, and improves carbon sequestration.
- **Agroforestry:** Integrating trees into farming systems can increase carbon storage in both the soil and vegetation while improving biodiversity and providing additional income sources.
- **Managed Grazing:** Rotating livestock through grazing areas allows pastures to regenerate, improving soil health and increasing carbon storage.

Benefits of Carbon Farming:

- **Revenue Generation:** Farmers who adopt carbon sequestration practices **may be eligible to sell carbon credits to businesses and governments** seeking to offset their emissions. This offers an additional income stream for farmers.
- **Soil Health Improvement:** Many carbon farming practices, such as agroforestry and cover cropping, enhance soil fertility, leading to long-term productivity gains and healthier ecosystems.

Through the adoption of climate-smart agriculture and carbon farming, farmers can contribute to global climate change mitigation efforts while simultaneously improving their productivity, resource efficiency, and overall farm sustainability.

These technologies and practices are not just theoretical solutions; they are actively transforming how agriculture is practiced globally. By integrating precision farming, AI, blockchain, and climate-smart strategies, agricultural projects can adapt to the challenges of the 21st century and contribute to a more sustainable, profitable, and resilient agricultural landscape.

13.4 Vertical and Urban Farming

As urbanization continues to increase globally, the demand for food in cities is also rising. However, arable land is becoming scarce, which creates challenges for traditional agriculture. Vertical farming and urban agriculture offer innovative solutions to these challenges by utilizing limited space within urban environments. These practices enable food production in stacked layers or in small, controlled environments like rooftops, warehouses, and even abandoned buildings.

13.4.1 Advantages of Vertical and Urban Farming

1. Space Efficiency:

Vertical farming maximizes the use of limited space in cities by growing crops in vertically stacked layers. This allows urban areas with limited land availability (such as rooftops, vacant lots, and indoor spaces) to be converted into productive food-growing zones. For instance, a single floor of a building can host multiple levels of crops, significantly increasing the area available for farming compared to traditional horizontal farming.

2. Reduced Transportation Costs:

By growing food closer to the point of consumption, vertical farming reduces the need for long-distance transportation, which in turn lowers carbon emissions and transportation costs. This not only makes urban farming more economically viable but also reduces the environmental footprint of food production. Fresh produce grown locally can be delivered faster, ensuring higher quality and less spoilage during transportation.

4. Year-Round Production:

Unlike traditional farming that is dependent on seasons, indoor vertical farms can produce crops year-round. Controlled indoor environments regulate factors like temperature, humidity, and light, making it possible to grow crops independent of weather conditions. This ensures a consistent food supply, even in areas with extreme weather, contributing to food security.

5. Reduced Environmental Impact:

Vertical farming uses fewer pesticides and herbicides compared to conventional farming, which helps protect surrounding ecosystems. Additionally, indoor farming reduces the need for large expanses of land, mitigating the risk of deforestation and land degradation.

13.4.2 Key Technologies in Vertical Farming

1. Hydroponics:

Hydroponics is a soil-less growing method in which plants are cultivated in a nutrient-rich water solution. This method is highly water-efficient since the water can be recycled and reused, significantly reducing water consumption compared to traditional farming. Hydroponics also provides precise control over nutrient levels, leading to healthier plants and higher yields.

2. Aeroponics:

Aeroponics is a technique where plants are grown with their roots suspended in the air, and nutrients are delivered through a fine mist. This method is even more water-efficient than hydroponics and allows for greater control over the growing environment. Since plants in aeroponic systems receive more oxygen, they tend to grow faster and more robustly.

3. LED Lighting:

LED lights are specially designed to mimic the sunlight spectrum, enabling plants to perform photosynthesis in indoor settings. By controlling light cycles with LED lighting, farmers can optimize growth conditions, boost yields, and reduce energy consumption. Full-spectrum LED lights can be used to tailor light to specific crops, enhancing their growth at different stages (e.g., seedling, vegetative, flowering).

13.5 Automation and Robotics in Agriculture

As labor shortages and increasing demand for efficiency become more prevalent in agriculture, automation and robotics are set to play a pivotal role in transforming farming practices. The introduction of robotic systems can optimize various agricultural processes, including planting, weeding, harvesting, and monitoring.

13.5.1 Robotic Harvesting

Robotic harvesters are designed to pick fruits and vegetables with precision, ensuring that the produce is not damaged during harvesting. This is particularly beneficial for delicate crops such as berries, which are often prone to bruising. By automating this process, farmers can reduce labor costs and mitigate the challenges associated with labor shortages.

- **Improved Quality:** Automated harvesting machines are capable of picking crops at the perfect ripeness, ensuring better quality produce. This reduces waste caused by overripe or underripe crops that may not meet market standards.
- **Cost Reduction:** In regions where labor costs are high, robotic harvesters can significantly reduce the need for manual labor, thereby lowering operational expenses. These robots can work around the clock, leading to faster harvests and less spoilage.

13.5.2 Autonomous Tractors and Drones

Autonomous vehicles, including tractors, planters, and harvesters, are equipped with AI to perform essential farming tasks without human intervention. These machines can autonomously plow fields, plant seeds, apply fertilizers, and even harvest crops, all while optimizing for efficiency.

- **Increased Efficiency:** Autonomous machines can operate 24/7, completing tasks faster and more consistently than human workers. This leads to improved productivity, and since these machines are often GPS-guided, they can perform tasks with precision, minimizing waste and overlap.

- **Labor Savings:** By automating repetitive, labor-intensive tasks, farmers can allocate human resources to more specialized roles, such as managing operations, conducting research, or overseeing quality control.

Drones have become invaluable tools for aerial crop monitoring, precision spraying, and pest detection. Equipped with cameras and sensors, drones provide farmers with detailed insights into the health of crops, soil conditions, and potential pest outbreaks.

13.6 Policy and Regulatory Developments

The agricultural industry is heavily influenced by government policies, which can either encourage or discourage certain farming practices. As climate change concerns intensify and the global food system becomes increasingly complex, governments are implementing new regulations to ensure food security, sustainability, and climate resilience.

13.6.1 Incentives for Sustainable Practices

Governments may provide financial incentives such as subsidies, grants, or tax breaks to promote the adoption of sustainable farming practices. This includes support for organic farming, water conservation, soil restoration, and the reduction of greenhouse gas emissions. These incentives make it easier for farmers to transition to more sustainable methods by offsetting some of the costs associated with implementing new technologies or switching to environmentally friendly practices.

- Carbon Sequestration: Governments may also offer incentives for farmers to implement carbon farming practices, which sequester carbon in the soil, potentially generating carbon credits that can be sold to companies seeking to offset their emissions.

13.6.2 International Trade and Agri-Food Systems

Agricultural projects are also impacted by international trade policies that shape the movement of goods across borders. Policies affecting tariffs, market access, food safety standards, and sustainability certifications can significantly influence the profitability and logistics of agricultural ventures.

- **Trade Agreements:** Trade policies and agreements influence access to international markets, shaping how agricultural products are traded globally. Changes in these policies can affect pricing, supply chains, and the availability of products.
- **Food Safety and Sustainability Standards:** Global trends toward increased food traceability and sustainability mean that agricultural products must adhere to stringent standards to be competitive in international markets. This will influence farming practices worldwide, as producers will need to comply with emerging regulations around food safety and environmental responsibility.

13.7 Conclusion

The future of agricultural project management will increasingly revolve around the integration of technological innovations, sustainability, and resilience. From the rise of vertical farming in urban spaces to the use of automation and robotics on the farm, the agricultural industry is evolving to meet the demands of a growing global population in the face of climate challenges.

To stay competitive and ensure the success of agricultural projects, project managers must be proactive in adopting new technologies, staying informed about policy developments, and incorporating sustainability into their project designs. By doing so, they will contribute to building a secure, efficient, and sustainable agricultural future that can feed the world while minimizing environmental impact.

Chapter 14: Key Insights and Future Directions in Agricultural Project Management

Benefits:

This chapter offers students and researchers a practical understanding of agricultural project implementation through detailed real-world examples. By analyzing project successes and challenges, it enhances their ability to apply concepts, develop solutions, and avoid common pitfalls in future agricultural project planning and execution

14.1 Summary of Key Concepts

This book has taken a comprehensive journey through the various aspects of agricultural project management and feasibility studies, providing a detailed framework for managing agricultural initiatives effectively. We have covered both foundational principles and practical applications that are crucial for the successful planning, implementation, and sustainability of agricultural projects. Here are some of the key themes and concepts from the book:

1. **Agricultural Project Management Framework**

   The life cycle of an agricultural project, from initiation to closure, was thoroughly discussed. We explored how to effectively manage each stage, emphasizing the importance of:

   - **Initiation**: Defining the project's scope and objectives.
   - **Planning**: Creating detailed roadmaps, resource plans, timelines, and budgets.
   - **Execution**: Implementing the plans and ensuring tasks are completed.
   - **Monitoring**: Tracking progress, identifying issues early, and adapting as necessary.
   - **Closure**: Finalizing the project, conducting evaluations, and ensuring sustainability.

2. **Feasibility Study**

   The feasibility study is essential to understanding whether an agricultural project can succeed. We looked at various aspects of feasibility, including:

   - **Market Feasibility**: Analyzing demand, competition, and market trends.
   - **Technical Feasibility**: Assessing the practicality of technology and infrastructure for the project.
   - **Financial Feasibility**: Ensuring the project's financial viability and potential return on investment.
   - **Social and Environmental Feasibility**: Evaluating the potential impacts on communities and ecosystems, ensuring the project aligns with sustainability goals.

3. **Planning and Implementation**

   The book provided practical guidance on how to manage resources, mitigate risks, and track progress. This includes:

   - **Resource Management**: Optimizing the use of land, labor, and capital.
   - **Risk Mitigation**: Identifying potential risks early on and creating contingency plans.
   - **Monitoring and Evaluation (M&E)**: Establishing systems to track performance and make adjustments when needed.

4. **Case Studies and Best Practices**

   We examined real-world agricultural projects across different scales and sectors. These case studies offered valuable lessons on:

   - **Managing smallholder cooperatives**.
   - **Scaling up large agribusiness ventures**.
   - **Implementing sustainable farming practices** in challenging environments. These examples helped illustrate how effective

management, coupled with sound practices, leads to enhanced productivity, sustainability, and economic growth.

5. **Emerging Trends**

   The book also explored the future of agricultural project management, focusing on innovations such as:

   - **Precision agriculture**: Leveraging data to optimize farming practices.
   - **Artificial Intelligence (AI) and Machine Learning (ML)**: Enhancing decision-making and predictive analytics.
   - **Vertical farming**: Maximizing space efficiency in urban areas.
   - **Climate-smart agriculture**: Implementing practices that are resilient to climate change while reducing environmental footprints.

14.2 The Importance of Sound Project Management in Agriculture

Effective agricultural project management is more than just ensuring a high yield. It involves creating a balance between economic success, environmental sustainability, and social responsibility. Proper project management ensures:

- **Optimal use of resources**: By managing finances, labor, and materials efficiently, projects are more likely to remain on budget and meet their objectives.
- **Risk minimization**: Well-planned and well-executed agricultural projects can identify potential risks and mitigate them before they escalate.
- **Alignment with project objectives**: A structured approach ensures that the outcomes align with both the short-term and long-term goals of the project.

Multidisciplinary Approach

Given the complexity of modern agriculture, project managers need expertise across various domains:

- **Project Management**: Planning, organizing, and overseeing the entire project life cycle.
- **Finance**: Ensuring financial health and profitability through budgeting, cost management, and financial forecasting.
- **Environmental Science**: Implementing sustainable practices that minimize ecological impacts.
- **Social Development**: Understanding community needs and ensuring that the project benefits local populations and adheres to social responsibility principles.

Agricultural project managers must also navigate the local challenges—such as land availability, labor force issues, and regional policies—while adapting to global market forces, like fluctuating commodity prices and trade regulations. Effective management practices will ultimately help future-proof agricultural projects, ensuring they are resilient, profitable, and impactful in the face of these ongoing challenges.

14.3 The Role of Feasibility Studies in Project Success

One of the most important lessons from this book is the critical role of feasibility studies in the success of agricultural projects. A feasibility study serves as a thorough evaluation of the viability of a project before significant investments are made, helping to ensure that projects are not only achievable but also financially sound.

Key Functions of Feasibility Studies:

1. **Identifying Risks Early**

Feasibility studies identify potential risks such as market volatility, technological challenges, environmental hazards, and social impacts. By spotting these risks early, project managers can take proactive measures to minimize or mitigate them.

2. **Ensuring Financial Viability**

   A comprehensive feasibility study provides insights into whether the project will generate the expected return on investment (ROI). It allows investors and stakeholders to make informed decisions about whether to move forward, ensuring that the financial structure is solid and that costs will not exceed projected earnings.

3. **Understanding Market Dynamics**

   Feasibility studies help assess demand, competition, consumer preferences, and distribution channels. Understanding these market dynamics ensures that the agricultural products meet demand and can be sold at a profit. This market analysis is crucial in avoiding overproduction or investing in crops that will not yield a competitive return.

4. **Assessing Environmental and Social Impacts**

   Feasibility studies must include evaluations of how the project will impact the environment and local communities. Projects that negatively affect the ecosystem or displace local populations risk long-term failure or opposition. A feasibility study identifies these issues and suggests measures for mitigating adverse effects, aligning the project with broader sustainability goals.

A well-prepared feasibility study is more than just a checklist—it is a strategic tool that guides decisions, ensuring that agricultural projects are feasible, sustainable, and aligned with broader objectives. By using this tool effectively, project managers gain the confidence needed to move forward with their projects, ensuring that resources are allocated efficiently and goals are met.

Final Thoughts

Through this book, we've learned that agricultural projects, especially in the context of modern challenges such as climate change, population growth, and food security, require careful management, innovative solutions, and a solid foundation based on feasibility. By following the principles of sound project management, conducting thorough feasibility studies, and integrating new technologies and sustainable practices, agricultural projects can not only succeed but also contribute positively to the broader goals of sustainability and food security.

14.4 Future Directions for Agricultural Project Management

The agricultural sector is experiencing profound transformations, driven by technological advancements, evolving consumer preferences, and increasing environmental pressures. To navigate this evolving landscape, agricultural project managers must be prepared to adapt and embrace the changes that lie ahead.

Here are some **key directions** for future agricultural project management:

1. **Embracing Technology and Innovation**

   The rapid pace of technological innovation, particularly in artificial intelligence (AI), drones, blockchain, and vertical farming, is set to revolutionize agricultural project management. By leveraging these technologies, project managers can:

   - Enhance operational efficiency through automation and data-driven decision-making.
   - Improve resource allocation with the help of precision agriculture tools, such as GPS and IoT sensors.
   - Boost sustainability through smart farming practices, like AI-powered climate monitoring or automated irrigation systems that optimize water usage.

As agricultural project managers, it will be crucial to stay ahead of these technological trends, continuously upgrading skills and incorporating cutting-edge technologies into projects to improve outcomes and reduce costs.

2. **Sustainability as a Core Principle**

   As climate change continues to accelerate, sustainability will become a central focus in agricultural projects. Projects must integrate climate-smart practices that not only mitigate the environmental impacts of agriculture but also enhance resilience to climate risks. This could include:

   - **Water conservation**: Implementing rainwater harvesting or precision irrigation systems to optimize water use.
   - **Soil health**: Emphasizing practices like no-till farming, cover cropping, and agroforestry to improve soil structure and reduce erosion.
   - **Waste reduction**: Minimizing food waste through improved supply chains and creating circular systems where agricultural byproducts are repurposed for other uses.

   Failure to prioritize sustainability could result in reputational damage, regulatory hurdles, and long-term challenges in maintaining viable agricultural operations. For these reasons, project managers must champion sustainable development throughout the entire project lifecycle.

3. **Inclusive and Community-Centered Approaches**

   Agricultural projects, particularly those focusing on smallholder farmers or rural communities, must adopt inclusive development strategies. These projects should:

   - Empower local communities by promoting gender equality and offering opportunities for youth and marginalized groups.

- Provide education and training to ensure community members can take full advantage of the project's benefits.
- Encourage local ownership and participation in decision-making processes to ensure that the benefits of the project are shared widely.

The success of agricultural projects often hinges on how well they integrate the needs and aspirations of local communities. Therefore, community-centered approaches will be key to creating socially responsible and sustainable agricultural initiatives.

4. **Adapting to Changing Global Markets**

   The future of agricultural project management will be shaped by dynamic shifts in global markets. Changes in consumer behavior, emerging trade agreements, and increasing demand for sustainable products will influence how agricultural projects are designed and implemented. Key considerations include:

   - **Market demands**: As consumers increasingly seek sustainable and ethically produced food, projects must align with these preferences.
   - **Trade policies**: Global trade agreements and tariffs will affect access to markets, especially for export-oriented agricultural projects.
   - **Supply chain resilience**: Project managers will need to focus on building supply chain resilience in response to fluctuations in demand or disruptions caused by events like the COVID-19 pandemic.

Understanding and adapting to these global trends will be crucial for maintaining a competitive edge and ensuring that agricultural projects are not only economically viable but also aligned with the broader economic and political environment.

14.5 The Challenges Ahead

While the future of agricultural project management holds vast potential, it also presents significant challenges. These challenges require innovative solutions, adaptive strategies, and proactive management.

Key Challenges:

1. **Climate Change**

   Rising temperatures, erratic rainfall patterns, and more frequent extreme weather events will increase unpredictability in farming systems. Agricultural project managers must be prepared to:

   - Develop resilience strategies that enable projects to adapt to changing weather conditions.
   - Incorporate climate adaptation measures, such as drought-resistant crops or early-warning systems to anticipate and mitigate climate risks.

2. **Resource Scarcity**

   As the global population continues to grow, access to essential resources like land, water, and fertilizers will become more limited. Project managers must prioritize resource efficiency by:

   - Implementing water-efficient irrigation systems and focusing on sustainable water management practices.
   - Enhancing soil fertility through sustainable agricultural practices such as organic farming and soil regeneration techniques.
   - Exploring alternatives to traditional fertilizers to reduce dependency on non-renewable resources.

3. **Financial Constraints**

   Securing adequate funding for agricultural projects, especially in developing regions, can be a major obstacle. Project managers will need to adopt innovative financing models such as:

   - Impact investing, where investors seek both financial returns and positive social or environmental outcomes.
   - Crowdfunding or public-private partnerships to pool resources and share the financial burden.

   Financial innovation will be crucial in making agricultural projects economically viable, particularly for smallholder farmers or those in regions with limited access to traditional funding sources.

4. **Political and Regulatory Challenges**

   Agriculture is highly influenced by **political stability**, **regulatory policies**, and trade agreements. Changes in government policies, international trade regulations, or agricultural subsidies can significantly impact project outcomes. Agricultural project managers will need to:

   - Stay informed about policy shifts and trade regulations that may affect their projects.
   - Ensure compliance with local, national, and international regulations, particularly regarding sustainability, food safety, and labor rights.
   - Build flexibility into project plans to adjust to changing political environments.

14.6 Final Thoughts

Agricultural project management is a dynamic and ever-evolving field. The integration of technological advancements, the growing importance of sustainability, and the push for inclusive development will define the success of agricultural projects in the coming decades.

Project managers will play a pivotal role in shaping the future of global food production. By staying ahead of technological trends, embracing sustainable practices, and ensuring the equitable distribution of benefits, agricultural project managers can help create projects that are not only profitable but also socially and environmentally responsible.

As the agricultural sector continues to evolve, the principles outlined in this book will remain as relevant as ever. Agricultural project managers must be equipped with both practical knowledge and visionary thinking to navigate challenges and capitalize on opportunities, ultimately contributing to a more sustainable, secure, and food-resilient future.

14.7 A Call to Action

As you conclude this journey through agricultural project management, I encourage you to apply what you've learned in real-world projects. Whether you're a project manager, farmer, investor, or policy maker, the lessons outlined in this book offer a strong foundation for informed decision-making and creating impactful agricultural projects that contribute to a sustainable and food-secure future.

By embracing the insights and strategies shared in these chapters, you can be part of the transformative process in the agricultural sector and help shape a future that balances food production with environmental stewardship and social equity.

Appendices

Appendix A: Sample Feasibility Study Report Template

A feasibility study is an essential tool for evaluating the viability of an agricultural project before it is executed. It provides critical insights into whether a project is technically, financially, and socially viable. The template below is designed to guide you through the process of writing a comprehensive feasibility study report for an agricultural project. This document can be tailored to meet the unique needs of your specific project.

1. Executive Summary

The Executive Summary is a concise summary of the entire feasibility study, providing a snapshot of the key findings and recommendations. It is typically written last but placed at the beginning of the report to give readers a quick overview.

- **Overview of the project**: Provide a brief description of the agricultural project, including its purpose, location, and scope. For example, if the project involves establishing a sustainable organic farm, mention the farm's size, the crops to be grown, and the project's objectives.
- **Brief explanation of feasibility findings**: Summarize the results of the feasibility study, highlighting the major findings, such as the project's viability, potential challenges, and risks. This section should give the reader a quick understanding of whether the project is feasible or not.
- **Key recommendations**: Offer recommendations based on the findings. For instance, if the study suggests that the land is unsuitable, recommend alternative locations or adjustments to the project plan.

2. Project Background and Objectives

This section provides more in-depth information about the project's foundation, including its goals and the context in which it will operate.

- **Description of the project and its purpose**: Explain the core concept of the project, its goals, and why it is being undertaken. This could include details about the type of agricultural project (e.g., crop farming, livestock management, agroforestry) and the motivations behind it (e.g., food security, sustainability, economic growth).
- **Clear goals and objectives**: Clearly define the project's objectives. These should be measurable and time-bound. For example:
  - To establish a 100-acre organic farm producing three key crops within two years.
  - To create 50 local jobs in the first year of operation.
- **Background information on the agricultural context**: Provide context regarding the agricultural sector in the region. This might include relevant data about local farming practices, climate, and soil conditions, as well as existing infrastructure and resources.

3. Market Feasibility

Market feasibility assesses the demand and competition for the agricultural product(s), providing insights into the project's potential profitability.

- **Market Demand Analysis**: Evaluate the demand for the product(s) you intend to produce. This could involve gathering data on consumption patterns, market size, and growth projections. For example, if the project involves organic vegetables, research into the growing demand for organic produce in the region will be essential.
- **Market Trends**: Identify key trends in consumer preferences, price fluctuations, and other relevant market dynamics. Are consumers increasingly looking for sustainable or locally-produced foods? What is the

forecast for agricultural commodity prices? These trends should be analyzed to understand the long-term potential of the market.

- **Competitor Analysis**: Conduct a thorough analysis of the competitive landscape. Identify existing competitors, their strengths and weaknesses, and the market share they control. This analysis will help you assess whether there's room for your project in the market and what unique value proposition you can offer.

4. Technical Feasibility

This section focuses on the practical aspects of implementing the agricultural project, such as land suitability, technology requirements, and resource availability.

- **Land and Soil Analysis**: Analyze the suitability of the proposed land for agricultural activities. This may include soil testing, assessing land topography, and determining whether the soil is conducive to the chosen crops or farming practices.
- **Water Resources**: Evaluate the availability and efficiency of water for irrigation. If your project requires significant water usage, investigate the proximity to water sources, water rights, and the efficiency of irrigation systems (e.g., drip irrigation vs. traditional sprinkler systems). If water resources are scarce, consider alternatives such as rainwater harvesting or water recycling.
- **Technology Requirements**: Identify the technology and equipment needed for the project. This could include advanced irrigation systems, climate control technologies for greenhouses, or agricultural machinery for planting and harvesting. Assess the feasibility of acquiring, maintaining, and operating these technologies.

5. Financial Feasibility

Financial feasibility assesses whether the project is financially viable and sustainable in the long term.

- **Cost Estimation**: Provide a detailed breakdown of the costs involved in the project. This should include:
  - **Fixed costs**: Land acquisition, infrastructure, machinery, and technology.
  - **Variable costs**: Labor, seeds, fertilizers, irrigation, and other ongoing expenses.
- **Revenue Projections**: Estimate the expected income from the agricultural project. This will include sales revenue based on production forecasts, market prices, and demand. Provide projections for the first few years, considering growth potential.
- **Profitability Analysis**: Perform financial analyses such as:
  - **Break-even point**: Calculate when the project will start generating a profit after covering all costs.
  - **Return on Investment (ROI)**: Estimate the ROI based on projected revenues and costs.
  - **Financial risks**: Identify potential financial risks, such as market volatility, climate-related losses, or changes in consumer preferences, and how they will be mitigated.

6. Social and Environmental Feasibility

This section addresses the broader impacts of the project on society and the environment, ensuring that the project contributes positively to local communities and the ecosystem.

- **Social Impact**: Evaluate the project's potential effects on local communities. This includes the creation of jobs, improvements in local infrastructure, and the impact on local economies. Consider any social challenges, such as potential displacement of existing communities or changes in land use that could affect local populations.

- **Environmental Impact**: Assess the environmental implications of the project, both positive and negative. This could include the impact on biodiversity, soil quality, water resources, and waste management practices. Consider sustainable farming practices, such as agroforestry or organic farming, that could mitigate negative environmental effects. Include recommendations for environmental monitoring and management.

7. Conclusion

The conclusion of the feasibility study summarizes the key findings and provides final recommendations.

- **Summary of feasibility findings**: Recap the overall feasibility of the project based on the assessments in previous sections. Was the project found to be technically, financially, and socially viable?
- **Recommended course of action**: Provide clear recommendations based on the findings. For example, if the project is deemed feasible, suggest the next steps for implementation, such as securing financing, acquiring land, or hiring staff. If the project is not feasible, suggest adjustments or alternatives.
- **Risk mitigation strategies**: Highlight potential risks identified in the study and propose strategies to mitigate them. This could include diversifying crops to reduce dependency on one product, implementing water conservation techniques, or seeking alternative financing options.

Appendix B: Financial Analysis Tools (Excel Templates, Formulas)

Financial analysis is crucial for understanding the economic viability of an agricultural project. By using the right tools and formulas, project managers can assess costs, revenues, and profits, which are essential for making informed decisions and securing financing. Below are some of the key financial analysis tools that can help you conduct detailed financial assessments for agricultural projects.

1. Cost Estimation Spreadsheet

A Cost Estimation Spreadsheet helps track and calculate both fixed and variable costs. These costs are essential to understand when calculating profitability and determining the overall financial feasibility of a project.

- **Fixed Costs** are those that do not change regardless of the level of production. These include costs like land acquisition, infrastructure, and equipment.
- **Variable Costs** change depending on the level of production. These include labor, fertilizers, irrigation, seeds, and fuel.

Example Formulas:

- **Total Cost**:

  Total Cost=Fixed Costs+Variable Costs\text{Total Cost} = \text{Fixed Costs} + \text{Variable Costs}Total Cost=Fixed Costs+Variable Costs

  This formula helps you sum the total costs by adding the fixed costs to the variable costs.

- **Break-even Analysis**: This determines the point at which total revenue equals total costs, meaning the project is neither making a profit nor a loss.

It is an essential metric for understanding when the agricultural project will start generating profit.

Break-even Point=Fixed CostsUnit Price−Variable Cost per Unit\text{Break-even Point} = \frac{\text{Fixed Costs}}{\text{Unit Price} - \text{Variable Cost per Unit}}Break-even Point=Unit Price−Variable Cost per UnitFixed Costs

- **Fixed Costs**: The total of all fixed costs (e.g., land, buildings, machinery).
- **Unit Price**: The price at which each unit of the product is sold.
- **Variable Cost per Unit**: The cost incurred for producing one unit of the product.

This formula helps you calculate the number of units that need to be sold to cover the fixed costs. Anything beyond this number contributes to profit.

2. Revenue Projection Model

The Revenue Projection Model helps estimate the expected income from the agricultural project, based on forecasted crop yields or production volumes.

Key Formulas:

- **Revenue**:

  Revenue=Unit Price×Quantity Sold\text{Revenue} = \text{Unit Price} \times \text{Quantity Sold}Revenue=Unit Price×Quantity Sold

  - **Unit Price**: The selling price per unit (e.g., price per kilogram of a crop).
  - **Quantity Sold**: The expected amount of product to be sold in a given period.

This formula can be adjusted for seasonality and market conditions. For example, the price of crops may fluctuate based on demand or external market trends, and the expected production can vary depending on weather conditions and resource availability.

- **Seasonal Adjustments**: If your project has seasonal crops, you may need to adjust expected yields and prices for different months or seasons.

For instance, if your crop produces more during a certain period, you can break down the revenue projections by season (e.g., spring, summer, fall).

3. Return on Investment (ROI) Calculation

ROI is a critical metric for assessing the profitability and financial viability of an agricultural project. This formula provides insight into the potential return you can expect on your investments.

Formula for ROI:

ROI=(Net ProfitTotal Investment)×100\text{ROI} = \left(\frac{\text{Net Profit}}{\text{Total Investment}} \right) \times 100ROI=(Total InvestmentNet Profit)×100

Where:

- **Net Profit** is the total revenue minus the total costs (fixed and variable).
- **Total Investment** refers to the capital invested in the project, including both equity and any debt financing.

The **ROI** formula helps you understand how effectively the capital invested is generating returns. For example, if you invest $100,000 and the project generates $120,000 in revenue, with $90,000 in total costs, the net profit is $30,000. Using the formula, ROI would be:

ROI=(30,000100,000)×100=30%\text{ROI} = \left(\frac{30,000}{100,000} \right) \times 100 = 30\%ROI=(100,00030,000)×100=30%

An ROI of 30% means that for every dollar invested, the project generates 30 cents in profit.

4. Cash Flow Forecast

A Cash Flow Forecast helps to project the movement of cash into and out of the agricultural project over a specific period, typically monthly or quarterly. This tool is essential for ensuring that the project has enough liquidity to cover its operational costs and to identify potential financial challenges, especially during the initial stages of the project when cash flows may be tight.

Key Components of a Cash Flow Forecast:

- **Cash Inflows**: This section tracks all sources of income, such as revenue from crop sales, subsidies, grants, or loans.
- **Cash Outflows**: This section tracks all expenses, including fixed and variable costs, equipment purchases, interest on loans, and any unexpected expenses.

Formula for Cash Flow:

Net Cash Flow=Cash Inflows−Cash Outflows\text{Net Cash Flow} = \text{Cash Inflows} - \text{Cash Outflows}Net Cash Flow=Cash Inflows−Cash Outflows

For each period (e.g., monthly or quarterly), you will calculate:

- **Opening Balance**: The cash available at the beginning of the period.
- **Closing Balance**: The cash available at the end of the period, calculated by adding the net cash flow to the opening balance.

For example:

- **Opening Balance**: $10,000
- **Cash Inflows**: $15,000 (revenue from crop sales)
- **Cash Outflows**: $8,000 (operating costs)

Net Cash Flow=15,000−8,000=7,000\text{Net Cash Flow} = 15,000 - 8,000 = 7,000Net Cash Flow=15,000−8,000=7,000

So, the **Closing Balance** will be:

Closing Balance=10,000+7,000=17,000\text{Closing Balance} = 10,000 + 7,000 = 17,000Closing Balance=10,000+7,000=17,000

This allows project managers to ensure they have enough funds to meet their obligations, particularly during peak expenses like harvest time or when investments in new equipment are required.

Using These Tools in Excel

These tools can be implemented easily in Excel by setting up separate worksheets or templates for each of the financial analysis categories (cost estimation, revenue projections, ROI, and cash flow). Excel provides a user-friendly environment to organize, calculate, and visualize financial data.

- **Excel Functions**:
  - Use the **SUM** function to calculate totals, like the total costs and revenues.
  - Use **IF** and **IFERROR** functions to handle conditional calculations or to display error messages if data is missing.
  - Graphs and charts (such as pie charts or bar graphs) can be created in Excel to visualize cost breakdowns, cash flows, and revenue projections for better decision-making.

Appendix C: Glossary of Key Terms

This glossary provides definitions for essential terms used in agricultural project management and feasibility studies. Understanding these terms is fundamental to navigating the field of agricultural projects and making informed decisions. Below are the key terms that are crucial for professionals in agricultural project management.

Agribusiness

Agribusiness refers to the broad range of business activities involved in the production, processing, and distribution of agricultural products. This includes all stages of the agricultural supply chain, from farming and livestock production to the processing, marketing, and retail of agricultural goods. Agribusiness encompasses both small-scale enterprises and large corporations involved in agriculture and food production.

Break-even Analysis

Break-even Analysis is a financial calculation used to determine the point at which an agricultural project's total revenues equal its total costs, resulting in neither profit nor loss. This analysis helps project managers and investors understand how much of a product needs to be sold at a given price to cover both fixed and variable costs.

- **Formula**:

Break-even Point=Fixed CostsUnit Price−Variable Cost per Unit\text{Break-even Point} = \frac{\text{Fixed Costs}}{\text{Unit Price} - \text{Variable Cost per Unit}}Break-even Point=Unit Price−Variable Cost per UnitFixed Costs

It is a crucial tool for assessing financial viability, especially for determining the minimum performance required for a project to be sustainable.

Carbon Farming

Carbon Farming refers to agricultural practices that are specifically designed to capture and store carbon dioxide (CO_2) from the atmosphere in soil. This is part of the broader effort to mitigate climate change by reducing the amount of greenhouse gases in the air. Carbon farming techniques include practices like agroforestry, cover cropping, and reduced tillage, which enhance soil health and its ability to sequester carbon.

These practices contribute to sustainable agriculture by promoting environmental benefits such as soil fertility and climate resilience.

Climate-Smart Agriculture (CSA)

Climate-Smart Agriculture (CSA) is an approach to farming that seeks to increase agricultural productivity and farmers' resilience to the impacts of climate change while minimizing or reducing greenhouse gas emissions. CSA integrates three core goals:

1. **Sustainability**: Promoting sustainable land and water use.
2. **Adaptation**: Building resilience to climate change impacts, like droughts or floods.
3. **Mitigation**: Reducing emissions of greenhouse gases from agricultural activities.

CSA involves practices such as crop diversification, agroforestry, water conservation techniques, and sustainable soil management to address both immediate and long-term challenges posed by climate change.

Feasibility Study

A Feasibility Study is a comprehensive evaluation process used to assess the viability of a proposed agricultural project from multiple perspectives—technical, financial, social, and environmental. The goal of a feasibility study is to provide a clear understanding of the risks and opportunities associated with the project, helping decision-makers determine whether it is worth pursuing.

Key components include:

- **Market Feasibility**: Analysis of market demand, competition, and pricing.
- **Technical Feasibility**: Assessment of land, technology, infrastructure, and resources required.
- **Financial Feasibility**: Evaluation of projected costs, revenues, and profitability.
- **Social and Environmental Feasibility**: Assessment of the social impacts and environmental sustainability.

Market Feasibility

Market Feasibility is an assessment of whether a product or service will succeed in the market based on various factors, including demand, competition, pricing, and consumer trends. For agricultural projects, market feasibility is crucial in understanding whether the product will meet market needs and generate sufficient demand to be profitable.

It involves:

- **Demand Analysis**: Estimating the quantity of products that will be purchased by consumers.
- **Competition Analysis**: Identifying current competitors in the market and analyzing their strengths and weaknesses.
- **Pricing Strategy**: Determining the optimal price point for the product that aligns with market demand and cost considerations.

Precision Agriculture

Precision Agriculture involves using technology and data analytics to optimize crop production, improve resource efficiency, and minimize waste. This approach relies on tools such as GPS, sensors, and drones to monitor and manage the variability of field conditions, such as soil moisture, temperature, and nutrient levels. By applying precise amounts of water, fertilizers, and pesticides, farmers can reduce costs, increase yields, and minimize environmental impact.

Key components include:

- **GPS Technology**: For mapping and tracking farm activities.
- **Sensors**: For monitoring soil health, moisture, and crop health.
- **Data Analytics**: To interpret data for better decision-making.

Return on Investment (ROI)

Return on Investment (ROI) is a financial metric used to evaluate the profitability of an investment by comparing the net profit to the total cost of the project. It is a key indicator for assessing the financial success of an agricultural project or business. A higher ROI indicates a more profitable investment.

- **Formula**:

ROI=(Net ProfitTotal Investment)×100\text{ROI} = \left(\frac{\text{Net Profit}}{\text{Total Investment}} \right) \times 100ROI=(Total InvestmentNet Profit)×100

The ROI calculation helps investors and project managers understand whether their resources are being used effectively and whether the project will deliver expected returns.

Sustainability

Sustainability refers to the ability of an agricultural project or practice to operate in a way that does not deplete natural resources, thereby ensuring that the project can be maintained over the long term. Sustainable agricultural practices aim to protect and enhance environmental health, support economic growth, and ensure social equity. This involves efficient resource use, reducing environmental impact, and promoting biodiversity.

In agricultural projects, sustainability can be achieved through practices like organic farming, conservation tillage, water conservation, and agroecological methods.

Vertical Farming

Vertical Farming is the practice of growing crops in stacked layers or vertically inclined surfaces, often in controlled indoor environments such as warehouses or on rooftops in urban areas. This method maximizes land use by growing crops vertically rather than horizontally, making it particularly useful in cities with limited available land.

Vertical farming often uses technologies like hydroponics or aeroponics, which allow crops to be grown without soil, using water-based or mist-based nutrient solutions. It allows for year-round crop production in areas with harsh climates and reduces transportation costs by bringing food production closer to urban centers.

Understanding these key terms is essential for anyone involved in agricultural project management, whether you're a farmer, project manager, investor, or policymaker. Each of these concepts plays a crucial role in ensuring that agricultural projects are sustainable, financially viable, and capable of meeting market demand.

Appendix D: References and Further Reading

This section provides a list of authoritative references and further reading materials for individuals interested in deepening their knowledge of agricultural project management and feasibility studies. These resources cover a wide range of topics, from climate change and food security to sustainable farming practices and financial feasibility in agriculture.

1. **FAO (Food and Agriculture Organization of the United Nations). (2017). The State of Food and Agriculture 2017: Climate Change, Agriculture, and Food Security. FAO.**

   This report by the FAO explores the relationship between climate change and agriculture, offering a detailed analysis of the challenges facing the agricultural sector. It provides insights into how climate change is affecting food security and agriculture globally and recommends strategies for improving resilience in agricultural systems.

   - **Key topics covered**: Climate-smart agriculture, food security, adaptation strategies, and the role of agriculture in climate change mitigation.

2. **International Fund for Agricultural Development (IFAD). (2019). Rural Development and Climate Change: A Guide for Practitioners. IFAD.**

   This practical guide by IFAD outlines how rural development initiatives can be aligned with climate change mitigation and adaptation strategies. It provides recommendations for agricultural project managers and policymakers working in rural areas, with a focus on sustainable development, community resilience, and climate change.

   - **Key topics covered**: Rural development, community engagement, sustainable agricultural practices, and climate change adaptation.

3. **Smith, B. (2020). The Essentials of Agricultural Project Management. Routledge.**

Smith's book offers a comprehensive overview of agricultural project management, covering both theoretical concepts and practical applications. It discusses the key principles of project management, from initiation and planning to execution and closure, with a focus on the unique challenges of managing agricultural projects.

- **Key topics covered**: Project life cycle, risk management, stakeholder engagement, and project evaluation in the context of agriculture.

4. **National Agricultural Library. (2021). Sustainable Agriculture: Concepts, Principles, and Applications. U.S. Department of Agriculture.**

This resource from the USDA provides a deep dive into sustainable agricultural practices and principles. It discusses the importance of sustainability in farming and how various techniques can be applied to improve soil health, conserve water, and reduce environmental impacts.

- **Key topics covered**: Sustainable farming practices, soil conservation, water management, and biodiversity in agriculture.

5. **World Bank. (2021). Agriculture for Development: The World Bank's Strategy. World Bank Publications.**

This publication outlines the World Bank's approach to promoting agricultural development globally. It discusses key strategies for improving productivity, supporting smallholder farmers, and addressing the challenges of climate change, market access, and infrastructure in agricultural projects.

- **Key topics covered**: Agricultural development strategies, policy frameworks, market access, and the role of institutions in supporting agriculture.

6. **Hollands, B. (2018). Precision Agriculture: A Guide to Technology Integration for Farmers. Springer.**

This guide provides an in-depth look at the integration of modern technologies, such as sensors, drones, and data analytics, into agricultural practices. Hollands offers practical insights into how precision agriculture can improve productivity, reduce waste, and enhance decision-making.

- **Key topics covered**: Precision farming techniques, technology integration, data-driven decision-making, and resource efficiency in agriculture.

7. **Guyer, C. (2019). Financial Feasibility in Agriculture. Wiley-Blackwell.**

Guyer's book focuses on the financial aspects of agricultural project feasibility studies. It provides practical tools for assessing the financial viability of agricultural ventures, including cost estimations, revenue projections, and risk analysis.

- **Key topics covered**: Financial feasibility, cost estimation, revenue modeling, financial risk management, and profitability analysis.

8. **Johnson, S., & Brown, A. (2022). Sustainable Farming Practices and Management Strategies. Academic Press.**

Johnson and Brown explore sustainable farming practices and management strategies that are both economically viable and environmentally friendly. This book provides case studies and examples of successful sustainable farming practices in various regions.

- **Key topics covered**: Sustainable agriculture, management strategies, case studies, and the integration of environmental sustainability in farming operations.

9. **Agricultural Extension Services (AESA). (2020). Managing Agricultural Projects for Success: A Handbook for Project Managers. AESA.**

This handbook is an essential resource for agricultural project managers, offering practical guidance on how to effectively manage agricultural projects from inception to completion. It covers project planning, implementation, monitoring, and evaluation, with a focus on ensuring success in agricultural ventures.

- **Key topics covered**: Project management best practices, monitoring and evaluation, stakeholder engagement, and risk management.

Appendix E: Further Tools and Resources

For project managers and professionals seeking additional tools and resources, the following websites offer valuable information, industry insights, and networking opportunities:

• **AgFunder Network:**

AgFunder provides a platform for investment opportunities, market analysis, and industry news related to agri-tech startups. It connects investors with innovative projects and companies in the agricultural sector.

- **Website**: www.agfunder.com

• **Agri-Tech East:**

Agri-Tech East is a UK-based community of agri-tech innovators. The platform offers insights into new technologies, projects, and research developments that are transforming agricultural management and production systems.

- **Website**: www.agri-tech-east.co.uk

• **Project Management Institute (PMI):**

The Project Management Institute (PMI) offers certifications, best practices, and a range of tools for project managers in various industries, including agriculture. It provides access to learning resources, networking, and global project management standards.

- **Website**: www.pmi.org

• **Agricultural Finance and Risk Management Course (IFPRI):**

The International Food Policy Research Institute (IFPRI) offers an online course focused on agricultural financial management and risk mitigation strategies.

This course provides valuable tools for managing the financial aspects of agricultural projects.

- **Website**: www.ifpri.org

FSC
www.fsc.org
MIX
Papier | Fördert
gute Waldnutzung
FSC® C083411

Kolibri 360 GmbH
Ferdinand-Jühlke-Straße 7
99095 Erfurt, Deutschland
produktsicherheit@kolibri360.de

Druck:
CPI Druckdienstleistungen GmbH
im Auftrag der
Kolibri 360 GmbH
Ferdinand-Jühlke-Str. 7
99095 Erfurt